AC ELECTRONICS

A Step-by-Step Introduction

Heathkit/Zenith Educational Systems

A SPECTRUM BOOK

Prentice-Hall, Inc.   Englewood Cliffs, New Jersey 07632

Ron Lyon

Library of Congress Cataloging in Publication Data

Main entry under title:

AC electronics.

"A Spectrum Book."
Reproduces the text of the ed. published: Benton
Harbor, Mich. : Heath Co., ©1978; unit objectives and
programmed reviews omitted.
Includes index.
1. Electric circuits—Alternating current.   I. Heath-
kit/Zenith Educational Systems (Group)
TK1141.A3 1983   621.319'13   82-23166
ISBN 0-13-002121-0
ISBN 0-13-002113-X (pbk.)

This book is available at a special discount when ordered in bulk quantities. Contact
Prentice-Hall, Inc., General Publishing Division, Special Sales, Englewood Cliffs, N.J. 07632.

This work is adapted from a larger work entitled *AC Electronics* © 1981 Heath Company. Revised
Prentice-Hall edition © 1983.

A SPECTRUM BOOK

10   9   8   7   6   5   4   3   2   1

Printed in the United States of America

Manufacturing buyer Patrick Mahoney

ISBN 0-13-002121-0
ISBN 0-13-002113-X {PBK.}

Prentice-Hall International, Inc., *London*
Prentice-Hall of Australia Pty. Limited, *Sydney*
Prentice-Hall of Canada Inc., *Toronto*
Prentice-Hall of India Private Limited, *New Delhi*
Prentice-Hall of Japan, Inc., *Tokyo*
Prentice-Hall of Southeast Asia Pte. Ltd., *Singapore*
Whitehall Books Limited, *Wellington, New Zealand*
Editora Prentice-Hall do Brasil Ltda., *Rio de Janeiro*

CONTENTS

Unit 1

ALTERNATING CURRENT

INTRODUCTION

Any individual who plans to work in the electronics field as a technician, engineer, or scientist must be familiar with alternating current. The same is also true for electronic hobbyists and experimenters or anyone who works with electronic equipment or circuits.

Alternating current (commonly identified as AC or ac) plays an important role in electronics. In fact, it is used more extensively than direct current (dc) because it has a wider range of practical applications. Alternating current is used so extensively that it is almost impossible to avoid using it, or at least being affected by it. For example, each time you plug in your electric shaver, toaster, or drill, you are using alternating current. When you turn on your radio or television set, the music or voice information that you hear is produced by alternating current. Even when you drive down a city street, you are still under its influence since the traffic signals that you obey and the street lights which illuminate your way all rely on alternating current for operation.

In this unit you will learn why alternating current is more useful than direct current. You will see how the most basic type of alternating current is produced and you will examine its important electrical characteristics. You will also briefly examine other types of alternating current which are used.

THE IMPORTANCE OF AC

Since it has many desirable characteristics, alternating current (ac) is suitable for many applications. In fact, alternating current is the most widely used type of electricity and it plays a fundamental and important role in various commercial, industrial and military applications.

Anyone working in the electronics field must understand the basic concepts behind ac and must also realize the advantages that it has to offer as well as the various ways in which it can be used. You will therefore begin this unit by learning basically what ac is and also why and how it is used. This brief discussion will help you realize the significance of ac and prepare you for the more detailed discussions which will follow.

What is AC?

Unlike direct current (dc) which flows in only one direction, alternating current (ac) periodically changes its direction of flow. In other words, alternating current flows first in one direction and then in the opposite direction.

The difference between direct current and alternating current is illustrated in Figure 1-1. Figure 1-1A shows a resistor which has direct current flowing through it. This current flows in only one direction as shown and results because of the fixed or constant voltage applied to the resistor. The fixed voltage source (referred to as a dc voltage) could be a simple storage battery or dry cell. The dc voltage source and resistor form a simple dc circuit. The current through this circuit flows only in the direction shown, since the polarity of the dc voltage source remains fixed.

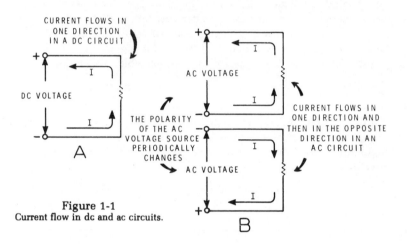

Figure 1-1
Current flow in dc and ac circuits.

Figure 1-1B shows how alternating current flows through a resistor. This current flows first in one direction and then in the opposite direction because the polarity of the voltage source changes as shown. This action is continuously repeated so that the current continually flows back and forth. The voltage source must therefore constantly reverse its output polarity in order to produce the resulting ac current. Such a voltage source is generally referred to as simply an ac voltage. The ac voltage source and resistor form a simple ac circuit.

4

It is also important to note that a direct current usually has a steady or constant value since it is usually produced by a dc voltage that has a fixed value. However, momentary changes may result if, for example, the dc voltage is adjusted to a higher or lower value or if the circuit resistance changes in value. However, in most dc circuits we are concerned with a steady current which always flows in one direction. By comparison, an alternating current usually changes in both value and direction. In other words, the current in an ac circuit will increase from zero to some maximum value and then drop back to zero as it flows in one direction and then vary in the same manner in the opposite direction. The exact manner in which the current increases and decreases in each direction can be controlled, thus making it possible to produce various types of ac signals. In fact, a variety of ac voltage sources are used to generate different types of ac voltages that are suitable for various applications. Later in this unit you will examine some of the basic types of ac signals that are used.

Why is AC Used?

Alternating current is widely used because of its versatility. Since ac changes in both value and direction, it has characteristics which can be used in a wide range of applications.

For example, when a large amount of electrical energy is required for a particular application, it is much easier to generate and transmit alternating current instead of direct current. In applications where large amounts of power are required, devices such as batteries (which produce dc voltages suitable for low power applications) cannot be used. In these applications, electromechanical devices known as *generators* (to be described later in this unit) are used to generate the high voltages and currents required. Although generators can be used to produce both dc and ac electricity, ac generators are less complex, they can be constructed in larger sizes, and they are often more economical to operate. Therefore, ac electrical power is simply easier and cheaper to produce.

An ac voltage can be easily transformed to a higher or lower voltage by passing it through a relatively simple device known as a *transformer*. Furthermore, this increase or decrease in voltage can be achieved with very little loss in power. In other words, the new voltage will provide approximately the same power to a load as could be obtained from the initial voltage. This is an important feature which is used to advantage in many applications. Although it is true that dc voltages may also be stepped up or down, the process is much more complex and costly. Also, considerable power is lost in the transformation of dc voltages thus making the conversion of dc voltages less efficient. This advantage, combined with the advantage previously described, makes ac highly suitable in applications where large amounts of electrical power are required.

Alternating current may be easily converted into direct current which can in turn be used to operate various types of dc circuits or equipment. Although it is true that dc power may be converted into ac power, the process is much more complex and is also more expensive and less efficient. Therefore, when ac is used as the primary source of electrical power, dc can still be obtained when needed by using a relatively simple conversion process. This feature gives ac another advantage over dc when used to provide electrical energy for the operation of electronic equipment or circuits.

The characteristics and features just described may seem to indicate that ac is useful only because it can serve as a source of electrical power which can be used to operate electronic equipment. However, this is not the case. Alternating current is also used extensively to transmit information from one location to another. This information carrying ability results because the characteristics of an alternating current or voltage can be made to vary in a desired manner. In other words, the magnitude or amplitude (maximum value in each direction) can be made to vary in a manner which will represent intelligence or information. Even the rate at which the alternating current changes direction can be made to vary in a desired manner and therefore represent intelligence. In this manner, information can be inserted within an alternating current or voltage, thus making it possible for the ac to carry information. When alternating currents or voltages are used to carry information they are often referred to as ac *signals*. AC signals are used extensively in electronics to carry information from one point to another within an electronic circuit. However, these signals can also be transmitted over long distances by using long wires or transmission lines.

Alternating current may also be converted into electromagnetic waves (also called radio waves) which can radiate or travel through space. This action is possible because a conductor which carries alternating current is surrounded by a magnetic field which expands and collapses as the intensity and direction of the current changes. If the current changes at a sufficiently high rate of speed, the magnetic field will actually radiate outward and the radiated energy will vary in accordance with the alternating current. This means that ac signals (which contain information) can be transmitted from one location to another without the use of wires or transmission lines. This action cannot be performed with direct current.

Although it is true that direct currents or voltages can be used to carry information, the dc signals produced are not as versatile as ac signals and do not have as many applications. This is because the amplitude of a direct current or voltage can be varied, but not the direction. In most cases, a dc signal is simply a series of pulses which are produced by interrupting a steady value of current or voltage. The information is therefore represented by a group of pulses in the form of a code.

The points just discussed illustrate just a few of the reasons why ac is used. Although there are many additional factors to consider, we can summarize by saying that ac is primarily used to either provide electrical power or to provide a means of transmitting information or intelligence from one point to another.

Where is AC Used?

Now that we have examined some of the basic reasons for using ac, we will consider some of its more important applications. To begin with, alternating current is used wherever a large amount of electrical power is required. In fact, almost all of the electrical energy supplied for domestic and commercial purposes is alternating current. The various electric power generating stations across the country produce alternating current which is used in various homes, factories and businesses. AC power is used because it is easy to generate and because it may be easily stepped up or down as described earlier.

7

Large amounts of ac power can be generated at the power plant by extremely large generators which are driven by turbines that are in turn powered by steam or falling water. Power stations that use falling water are referred to as *hydroelectric* stations and are usually located near a dam where the water can be stored and its rate of flow can be controlled. When steam is used in the generation process, a source of heat is required to produce the steam. This heat may be produced by burning coal, although many of the newer stations use nuclear energy.

The ac electrical power produced by a power plant is distributed as shown in Figure 1-2. The generators at the plant produce a relatively high ac voltage (often 2300 volts or more) and this voltage is passed through power transformers which step it up to an even higher voltage (often as high as 275,000 volts). This extremely high ac voltage is applied to long distance transmission lines which carry it to the various cities and towns that are serviced by the power plant. At each location where the power is to be used, it is passed through power transformers which step it back down to a lower voltage (typically 2300 volts). The ac voltage is then distributed through wires which are strung on utility poles that are located along roadways, streets, and alleys. This relatively high ac voltage may be directly used to power various types of high voltage motors and equipment used in various industries, but must be reduced even further before it can be used in the home. This final step-down is accomplished by a transformer which is usually located on one of the utility poles. This transformer steps the voltage down to 240 and 120 volts which is then distributed to the nearby homes.

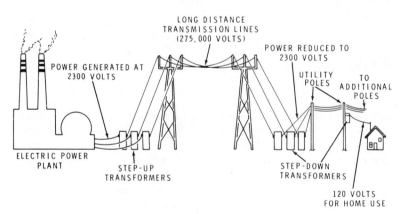

Figure 1-2
A typical electrical power distribution
system.

You are perhaps wondering why it is necessary to increase the ac voltage to such high values and reduce it again at the points where it is used. It is done so that the electrical energy can be transmitted with a minimum amount of loss through the transmission lines. When high voltages are used, the transmission lines are required to carry less current. With lower currents flowing through the transmission lines (which offer a specific amount of resistance), less power is dissipated by the lines in the form of heat. The same amount of electrical energy could be transmitted by using lower voltages and higher currents, but the higher currents would cause more power to be lost in the transmission lines. Also, the higher currents would require the use of heavier transmission line (larger diameter wire). Therefore, the expense is lower and the efficiency of transmission is greater when high voltages are used. Of course, the high voltages used in transmission are too dangerous to use in practical applications and must be reduced to usable levels. However, with ac power, this process can be accomplished quite easily as explained earlier.

In the home, ac power is used for heating, cooking, and lighting. It is used to operate applicances such as clothes dryers, refrigerators, electric ranges, microwave ovens, dishwashers, and vacuum cleaners. In fact, all of the large electrical appliances used in your home depend on the 240 or 120 volt ac power provided by the electric power company.

Many industries rely on ac power to operate the electric motors that drive their machinery. AC motors are used much more extensively than dc motors because of the readily available ac power. However, the ac motors are also less expensive, more rugged, and more efficient than comparable dc motors. In the relatively few applications where dc motors are required, such as the types used to operate elevators and certain machine tools, the ac power is converted to dc power, which is then used to drive the equipment.

In other industrial applications ac is used to heat various materials. In these applications, rapidly varying ac currents are allowed to flow through specially shaped coils of wire. The electromagnetic fields produced by the ac currents, will cause heat to be generated within any metal objects which are placed within the coils. This process is often used to heat-treat various metals. A similar process is also used in the medical profession where ac currents are used to produce heat within body tissue.

Without alternating current, radio or television communications would not be possible. This is because ac is used to carry the sound as well as the picture information that is transmitted from one location to another. A typical television broadcasting system is shown in Figure 1-3. The sound and picture information is converted into electrical signals by the microphone and television camera respectively. These signals are applied to the sound and picture transmitters, as shown, where they are used to vary the characteristics of the high power alternating currents which are produced by the transmitters. The resulting high power ac signals contain the sound and picture information and are applied to the transmitting antenna. The antenna converts the signals into electromagnetic waves and radiates the waves into space. When these waves intercept the antenna on a television set, they produce ac signals within the antenna that are identical to the signals applied to the transmitting antenna. These ac signals are separated and processed within the television set so that the sound and picture information is extracted and converted to an audible and visual display.

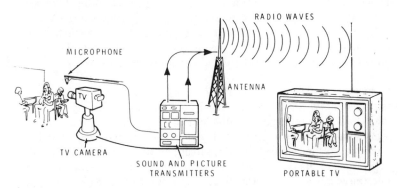

Figure 1-3

A typical television broadcast system.

AC signals also play an important role in many additional applications where information must be transmitted from one location to another. For example, the generation and transmission of ac signals are used in radar equipment, two-way radio communications systems, and also AM and FM radio broadcast systems. The various telephone systems across the country also use ac signals, although these signals are carried by wires or transmission lines.

In addition, ac signals are used in many short range applications where the distances between the transmitting and receiving points are only a few feet or possibly a few hundred yards. Such applications would include the radio control systems used to control model airplanes or boats, the remote television channel selectors which change the channels on your television set from several yards away, and the electronic garage door opener which allows you to open your garage door without getting out of your car.

As you can see, there are many applications for alternating current. In fact, there are so many that it would be impossible to consider all of them at this time. However, this brief review will help you to understand the important role that ac plays in the electronics field and some of the ways it is used to provide comfort, convenience, and even entertainment for a large number of people across the country.

GENERATING AC

Although alternating current may be generated in various ways the most basic means of obtaining ac is by using an electromechanical device known as an *ac generator* or *alternator*. An ac generator produces an alternating voltage which in turn will develop an alternating current through any load (resistor, lamps, etc.) that is connected to the generator's output terminals. Basically an ac generator produces an ac voltage by causing a loop of wire to turn within a magnetic field. This relative motion between the wire and the magnetic field causes a voltage to be induced within the wire. This voltage changes in magnitude and polarity as the speed and direction of the wire changes in relation to the magnetic field. The force required to turn the loop can be obtained from various sources. For example, the very large ac generators are turned by steam turbines or by the movement of water (hydroelectric) while the smaller units may be powered by gasoline engines.

We will now examine in detail the conditions which are necessary to produce a voltage within a conductor. Then we will see how this basic principle is used in ꞈ simple ac generator which is capable of producing an alternating voltage. Although this discussion will be very basic, it will provide you with beneficial background information.

Electromagnetic Induction

An ac generator is able to produce an alternating voltage because it makes use of a fundamental but important process known as *electromagnetic induction*. Electromagnetic induction is the process of inducing a voltage within a wire or conductor that is moving through a magnetic field. We will now examine in detail the conditions necessary for this process to occur.

The conditions necessary to produce electromagnetic induction are shown in Figure 1-4. Notice that a short wire or conductor has been inserted within the magnetic field that exists between the north (N) and south (S) poles of two magnets. When the conductor is moved upward through the magnetic field as shown, it cuts across the magnetic lines of force (also called magnetic flux) produced by the magnets. As the magnetic lines of force pass through the conductor, the free electrons in the conductor are forced to move. These free electrons will move towards one end of the conductor, thus leaving a deficiency of electrons at the other end. This excess and deficiency of electrons causes the opposite ends of the conductor to take on negative (−) and positive (+) charges as shown. In other words, a difference of potential or voltage is developed across the conductor.

The voltage developed across the conductor shown in Figure 1-4 results because of the relative motion between the conductor and the magnetic lines of force. This relative motion must exist in order for a voltage to be produced. The conductor may move while the field remains stationary or the conductor could be held stationary while the field is moved. Either condition would cause a voltage to be produced. If there is no relative motion, no voltage is produced.

The voltage produced within the conductor is generally referred to as an induced voltage. This voltage is induced within the conductor regardless of whether there is current flowing through the conductor or not. In fact a continuous current cannot flow through the conductor unless there is a complete circuit. Such a complete circuit might be formed by connecting a load (such as a resistor) across the conductor as shown by the dashed lines in Figure 1-4. Current could then flow through the load resistance as shown.

The polarity of the voltage induced within the moving conductor is determined by the direction that the conductor is moving and the direction of the magnetic field. Notice the direction of conductor motion and the direction (north to south) of the magnetic flux shown in Figure 1-4. Then note the polarity of the induced voltage and the direction that current would travel if a complete circuit was provided. If either the direction of conductor motion or the direction of the magnetic flux was reversed, the polarity of the induced voltage would be exactly opposite, and current would flow in the opposite direction.

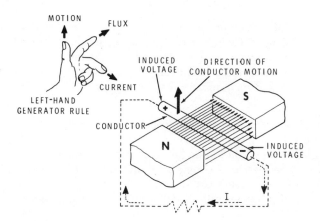

Figure 1-4
A voltage is induced within a conductor which moves through a magnetic field.

13

A simple rule can be used to determine the direction of current flow or induced voltage within a conductor. This rule requires the use of the thumb and first two fingers of your left hand and is commonly referred to as the *left-hand generator rule*. This rule is illustrated in Figure 1-4. When the left hand is positioned as shown, the thumb points in the direction of conductor motion, while the forefinger points in the direction of magnetic flux and the middle finger (which is bent out from the palm at 90°) points in the direction of induced current. If you examine the position of the fingers shown in Figure 1-4 and compare them to the conductor and magnetic field shown, you will find that the left hand rule verifies the fact that the induced current (assuming a complete circuit) will flow in the direction shown.

The amount of voltage induced in a conductor is determined by several factors which we will now consider. First of all, the induced voltage is affected by the strength of the magnetic field. A stronger magnetic field will result in more lines of force per unit area. This means that the conductor will cut more lines of force and the induced voltage will be higher. When the field strength is reduced, fewer lines of force exist and the induced voltage decreases.

The induced voltage also depends on the speed of conductor movement. The faster the conductor moves, the greater the induced voltage. This is because the faster moving conductor cuts more lines of force in a given period of time. When the speed of the conductor is reduced, fewer lines of force are cut in a given period of time and the induced voltage is lower.

The length of the conductor within the magnetic field also affects the induced voltage. The longer the conductor, the greater the induced voltage. This is simply because the longer conductor cuts more lines of force as it moves through the field. A shorter conductor intercepts fewer lines of force and its induced voltage must therefore be lower.

The angle at which the conductor cuts the magnetic field also affects the induced voltage. If the conductor moves at a right angle with respect to the field, as shown in Figure 1-4, the maximum amount of voltage is induced. However, as the angle between the field and the direction of conductor motion decreases, the induced voltage decreases. This relationship holds true regardless of the direction of the induced voltage.

14

The relationship between the direction of the field and the direction of conductor motion is shown in Figure 1-5. If the conductor (viewed edgewise) moves straight up from the starting position (direction A), it moves at a right angle (90°) with respect to the field. At this time the conductor is cutting the maximum number of lines per unit of time and the induced voltage is highest. Also, the voltage induced in the conductor would cause current to flow out of the page or toward you (according to the left-hand rule). The same condition also exists if the conductor moves in the opposite direction (direction E). In this case, the conductor is still moving at a right angle with respect to the field, thus allowing the highest voltage to be induced. The only difference is that the opposite direction of motion would cause the induced current to flow into the page instead of out of the page as before.

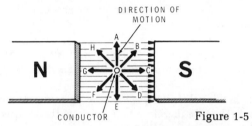

Figure 1-5

The induced voltage is determined by the rate at which the lines of force are cut.

If the conductor moves in direction B at the same speed, the cutting angle decreases below 90°. This means that the conductor must travel further between lines of force and fewer lines are cut for the same distance of travel. The induced voltage would therefore be lower. The same is also true if the conductor moves in direction H. In either direction B or H the cutting angle would be the same and the induced voltage would be the same. Also, in each case the induced current would flow out of the page. The same cutting angle could also be obtained if the conductor moved in direction D or F and the amount of induced voltage would likewise be the same. However, in direction D or F the induced current would flow into the page.

If the conductor moves in either direction C or G, the cutting angle is effectively zero. At this time the conductor moves parallel with the lines of force and no lines are cut. Under these conditions no voltage is induced in the conductor and no current can flow.

From previous discussion we can conclude that the amount of voltage induced in a conductor is affected by the following four factors:

1. The strength of the magnetic field.

2. The speed of conductor movement.

3. The length of the conductor in the field.

4. The angle at which the conductor cuts the field.

Although these four factors effectively state the various conditions which affect the voltage induced in a conductor, it is possible to formulate one simple rule which takes all of these factors into account. This simple rule is stated as follows:

The voltage induced in a conductor is directly proportional to the rate at which the conductor cuts the magnetic lines of force.

The word *rate* in this rule is used to indicate the number of lines of force cut per second. The rule therefore indicates that the induced voltage is proportional to the number of lines of force cut per second. When more lines of force are cut per second, the induced voltage increases and when fewer lines are cut per second, the voltage decreases. The number of lines of force cut per second (the rate) can be increased by increasing the strength of the field, the speed of conductor motion, the length of the conductor, or the cutting angle as previously explained.

A Simple AC Generator

Now that you are familiar with electromagnetic induction and the basic rules which govern this important process we will see how these rules can be used to construct a simple ac generator. Then we will see how the generator is used to produce an ac voltage.

Generator construction. A simple ac generator may be formed by bending a wire or conductor in the form of a loop and then mounting the loop so that it can rotate within a magnetic field. When the wire loop rotates, an ac voltage is induced within it. This induced voltage can be used to operate a load such as a resistor, lamp, or motor. The only other basic consideration is to provide a convenient means of extracting the ac voltage generated within the rotating loop and applying this voltage to the load. However, this problem can be easily solved as you will soon see.

A simple ac generator is shown in Figure 1-6. As mentioned previously, it basically consists of a wire loop called an *armature*, which is mounted so that it will rotate within the magnetic field produced between the north and south poles of a magnet as shown. The magnet used for this purpose is commonly referred to as a *field magnet*. The field magnet is constructed so that it produces a strong and evenly concentrated magnetic field between its poles. It can be either a permanent magnet or an·electromagnet. An electromagnet is preferred in applications where a high field strength is required in order to produce substantial output power.

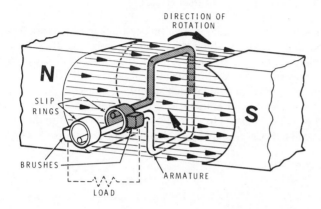

Figure 1-6
A basic ac generator.

The ac voltage that is induced within the rotating armature must be extracted at the ends of the wire loop which form the armature. However, the armature constantly turns, thus making it impossible to permanently attach any wires or leads (from a load) directly to the armature. For this reason it is necessary to use some type of sliding contact at each end of the wire loop. As you can see in Figure 1-6, two cylindrical metal rings are attached to the opposite ends of the wire loop. These metal rings are commonly referred to as *slip rings*. An external circuit or load is connected to these slip rings through contacts which are held against the rings. These contacts are usually made from a soft but highly conductive material such as carbon and are referred to as *brushes*. The brushes slide against the slip rings as the armature turns and in this manner the brushes serve as two stationary contacts to which an external load can be connected. The brushes essentially serve as the output terminals of the generator since the ac output voltage is applied through them to the load.

17

Generator operation. In order to function properly an ac generator must be operated so that its armature rotates at a constant speed. As the armature rotates in the magnetic field, the opposite sides of the armature always move in opposite directions. For example, one side may move down through the magnetic field while the other side moves up or vice versa. It is also important to note that in one complete revolution of the armature, each side must move down and then up through the field. Furthermore, each side of the armature always remains in contact with its respective brush (through a slip ring). Keeping these considerations in mind we will now examine the basic action that takes place throughout one complete revolution of the armature.

An armature is shown in four specific positions in Figure 1-7. These are intermediate positions which occur during one complete revolution of the armature. Notice that one side of the armature and its associated slip ring and brush are black while the other side is white. The two colors are used to help you keep track of each side of the armature. Also, a resistive load is connected to the brushes so that a complete circuit is formed which will allow current to flow through the armature and the load. The output voltage is monitored by a voltmeter which is connected across the load.

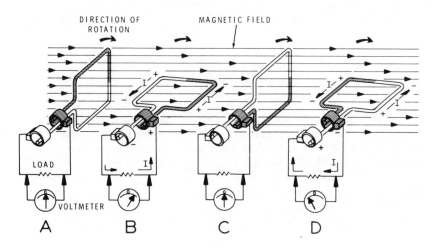

Figure 1-7
Generating an ac voltage.

Assume that the armature starts rotating in a clockwise direction from the initial position shown in Figure 1-7A. Notice that in this initial position, the black side of the armature is on top and the white side is on the bottom. As the armature moves from this starting position, the black side moves from left to right and the white side moves from right to left. However, both sides move parallel with the lines of force (no lines are cut) and therefore no voltage is induced in either side of the armature. The voltmeter, which is connected to the brushes, indicates zero at this time.

As the armature rotates from the position shown in Figure 1-7A, to the position shown in Figure 1-7B, the black side moves down through the field while the white side moves up. The opposite sides of the armature therefore cut the magnetic lines of force in opposite directions. This means that the polarity of the voltage induced in the black side will be opposite to the polarity of the voltage induced in the white side. However, the voltages induced in each side are actually series-aiding since the two sides of the armature form a loop. These induced voltages are equal in value and they enforce each other so that the resultant voltage which appears at the brushes is equal to the sum of the voltages induced in each side. The polarity of these voltages are shown in Figure 1-7B along with the resulting currents. Notice that the series-aiding voltages produce a current which circles through the armature and the load.

As you examine Figure 1-7B, notice that the armature is horizontal. In other words, the black and white sides of the armature are cutting the magnetic lines of force at right angles (the fastest cutting rate) resulting in the highest possible induced voltage. At this time the output voltage applied to the load is at its maximum value as indicated by the voltmeter. It is important to note that the output voltage did not suddenly jump from zero to maximum but increases at a specific rate. As the armature rotated from the position shown in Figure 1-7A to the position shown in Figure 1-7B, it cut the magnetic lines of force at a faster and faster rate until the maximum rate was obtained. This caused the output voltage to increase accordingly from zero to its maximum value.

When the armature rotates from the position shown in Figure 1-7B to the position shown in Figure 1-7C, it cuts the magnetic lines of force at a slower and slower rate. When the armature reaches the position shown in Figure 1-7C, the opposite sides of the armature are moving parallel to the lines of force so that no lines are cut. This means that the output voltage decreases from its maximum value to zero as the armature moves to the position shown in Figure 1-7C. The corresponding load current also decreases to zero along with the output voltage.

To this point the armature has completed one-half of a revolution and has produced an output voltage that has increased from zero to a maximum value and then decreased back to zero. The corresponding load current has also varied in the same manner. It is important to note that up to this point the output voltage has changed only in value and not in polarity or direction.

As the armature continues its rotation from the position shown in Figure 1-7C to the position shown in Figure 1-7D, the opposite sides of the armature again move across the magnetic lines of force in opposite directions. However, the black side of the armature now moves up and the white side moves down. This is exactly opposite to the situation which occurred during the first one-half revolution. Therefore, the voltages induced into each side of the armature have polarities which are opposite to those which were induced earlier. These induced voltages series-aid as before to produce a resultant output voltage. However, the output voltage now has a polarity which is exactly opposite to the polarity that was produced earlier. As the armature moves to the position shown in Figure 1-7D, the new output voltage (of opposite polarity) increases to maximum because the armature cuts the lines of force at the fastest rate at this position. This maximum voltage is indicated on the voltmeter which is connected across the load. The load current is also maximum at this time and it flows in the direction shown, which is opposite to its initial direction.

The armature completes a full revolution by returning to its initial position as shown in Figure 1-7A. As it continues its rotation and moves to this initial position, the armature cuts lines of force at a decreasing rate until no lines are cut at all. This causes the induced voltage within the armature and the resultant output voltage to decrease to zero. The corresponding load current also varies in the same manner.

As you can see, one complete revolution of the armature produces a voltage (and corresponding load current) that changes in both magnitude and direction. The output voltage and current are therefore ac values. However, this simple discussion has not fully explained exactly how these ac quantities vary. We have simply shown that an ac output voltage is produced.

We will now take a closer look at the basic generator action that was just described and consider the exact manner in which the ac output voltage is generated. We will again rotate the armature through the magnetic field but we will consider the action that takes place at a number of intermediate armature positions. Since the voltage induced in each side of the armature is equal and opposite (but also series-aiding), it is necessary to consider only the voltage that is induced in one side of the armature. Therefore, we will rotate only one side of the armature (one-half of the wire loop) through the magnetic field as shown in Figure 1-8. We will consider the action that takes place at 16 different armature positions (A through P) as shown. We will be looking into the end of the armature wire in much the same way that we viewed the conductor shown in Figure 1-5. Also, the armature wire will be rotated in a clockwise direction.

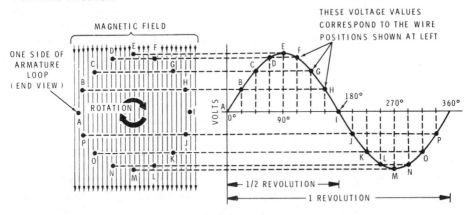

Figure 1-8
Plotting the ac output voltage.

At position A the wire moves parallel with the lines of force and no voltage is induced within it. However, as it moves to positions B, C, D, and then E, it cuts lines of force at a faster and faster rate. At position E the wire is moving at a right angle with respect to the field which results in the faster cutting rate. The voltage induced in the wire is proportional to the rate at which the lines of force are cut and since this cutting rate increases in a nonlinear manner, the voltage will also increase in the same way. This nonlinear increase in induced voltage is shown graphically in Figure 1-8. This voltage graph is plotted by marking off the angular rotation of the wire (in degrees) along a horizontal line. This horizontal line is calibrated for one complete revolution of the wire (360°). The voltage that is induced within the wire is plotted vertically as shown. Notice that five voltage values have been plotted (A through E) which correspond to wire positions A through E. Each successive value is higher and each value corresponds to the rate at which the magnetic lines of force are cut at the corresponding wire position. When these five values are connected to form a continuous line, we obtain a curve which shows how the induced voltage varies when the wire rotates one quarter of a revolution or 90°. Notice that the voltage increases rapidly from zero and then tapers off to a maximum value.

21

As the wire continues its rotation past position E and moves through positions F, G, H, and I, it cuts lines of force at a slower and slower rate thus causing the induced voltage to decrease accordingly. At position I, the wire moves parallel with the lines. At this time no lines are cut and the induced voltage is zero. The induced voltage values which correspond to wire positions F through I are plotted on the voltage graph as shown. Notice that these values (F through I) are successively lower, with the value at I being equal to zero. When these voltage values are joined with a continuous line we find that the voltage begins decreasing slowly from its maximum value and then it decreases at a faster and faster rate until it reaches zero. This portion of the voltage curve, between values E and I, shows how the voltage varies when the armature wire rotates an additional 90° or one-quarter of a revolution.

To this point, the armature wire has rotated one-half of a revolution or 180° and has produced a voltage that varies from zero to a maximum value and back to zero again. Now we will see what happens as the wire continues its rotation. When the wire moves past position I, it starts cutting lines of force again but this time it is moving in the opposite direction through the magnetic field. Therefore the polarity of the voltage induced in the armature wire will be opposite to the polarity of the voltage produced during the first one-half revolution. As the wire moves past position I and then through positions J, K, L, and M, it cuts lines of force at a faster and faster rate just as it did when it moved from position A to position E. Therefore, the voltage induced in the wire will increase just as it did before. The only difference is that it now has the opposite polarity. To show that the polarity is opposite, the voltage values which occur at positions J, K, L, and M are plotted below the horizontal line as shown. This portion of the curve shows that the voltage varies from zero to a maximum value in the opposite direction as the armature wire rotates an additional 90°.

When the armature wire moves on to positions N, O, and P, it cuts lines of force at a slower and slower rate. When it reaches position A (its initial starting position) it has completed one revolution and it is again moving parallel to the lines of force so that none are cut. The voltage therefore drops from its maximum value back to zero as shown. The voltage produced during the second half of rotation (between 180° and 360°) therefore increases from zero to a maximum value and drops back to zero again.

One complete revolution of the armature wire produces an ac voltage which varies in amplitude and direction as shown in Figure 1-8. During one-half of the revolution, the voltage increases from zero to maximum and drops to zero again. During the next half of a revolution the voltage varies in the same manner but has an opposite polarity. If this ac voltage is applied to a load, the resulting current through the load will vary in the same manner. In other words, the current will increase and decrease in one direction and then increase and decrease in the opposite direction.

Since the armature in an ac generator rotates at a constant speed, the ac output voltage produced by the device continually changes in value and direction as shown in Figure 1-8. Each complete revolution of the armature causes the output voltage to vary in the manner shown.

The ac generator just described represents the simplest device which could be used to generate an ac voltage. The ac generators that are used to produce electrical power for various commercial applications are more complex in construction, although they operate on the basic principles just described. Practical ac generators use many loops of wire within their armatures so that the induced voltage will be higher. They may also use more than one pair of north-south magnetic poles. This means that one complete revolution of the armature can produce more than one ac voltage variation.

Some ac generators are designed for relatively low power applications and are quite small. For example, the alternators that are now used on most automobiles are actually small ac generators (six or seven inches in diameter) which can produce only a few hundred watts of output power. These generators are powered by the car's engine and are used to produce an ac voltage which is converted to a dc voltage that is in turn used to operate the car's electrical system. The newer alternators are used in place of dc generators (which were once widely used) because they are more efficient and require less maintenance, even though it is necessary to convert their ac outputs into dc voltages.

Some ac generators are designed to produce large amounts of electrical power and are extremely large. For example, an ac generator that is used by an electric power company might be too large to fit into the living room of your house. Such a generator might produce as much as 1,000,000 watts of output power. This would be enough power for an entire community. These large generators are often turned by steam turbines or by the action of falling water as mentioned previously.

This extremely large ac generator, next to the man, is powered by the large turbines shown at the right. This generating system produces up to 609,000 kilowatts of output power. It is just one type of ac generator that is used by electric power companies to generate electrical power for the home and for industry.
(Courtesy of Westinghouse)

THE SINUSOIDAL WAVEFORM

In the previous section you learned how an ac voltage could be produced by a simple ac generator. This ac voltage was graphically plotted in Figure 1-8 so that you could see exactly how it varies throughout one complete revolution of the generator's armature. When voltage (or current) values are plotted in this manner and joined to form a continuous curve, they form a picture or pattern which is referred to as a *waveform*. The waveform shows exactly how the voltage varies over a period of time as the armature rotates. You may examine the waveform to determine the exact value and polarity of the ac voltage at a particular instant when the armature is at a specific point.

The waveform shown in Figure 1-8 varies in a unique manner and is given a special name. It is called a *sinusoidal waveform* or simply a *sine wave*. The sine wave is the most basic and widely used ac waveform. It can be produced by an ac generator as previously shown or it can be produced by various types of electronic circuits.

The ac voltage that you use in your home to provide heat, light, and operate appliances, varies in a sinusoidal manner. The radio and television signals which carry sound and picture information are basically sine waves whose characteristics have been modified or made to vary in a desired manner. Even various waveforms which are more complex than the simple sine wave can be proven to be mathematically equivalent to various combinations of sine waves.

Since the sinusoidal waveform is extremely important, we will examine it in greater detail. First, we will explain why it is called a sine wave and then we will consider its most important electrical characteristics.

The Basic Sine Wave

The sine wave is referred to as a *sine* wave because it changes in value according to the trigonometric function known as the *sine*. The sine is simply a trigonometric function which relates to either of the angles (except the right angle) in a right triangle. A right triangle is simply a triangle that has one right angle (a 90° angle). An angle has a sine value that is equal to the ratio of the length of the opposite side (the side opposite to the angle) to the length of the hypotenuse (the side opposite the right angle). This relationship is shown in Figure 1-9. Notice that the sine of angle A is equal to the opposite side (side Y) divided by the hypotenuse (side Z). Angle B would have a sine value that is equal to its opposite side (side X) divided by the hypotenuse (side Z).

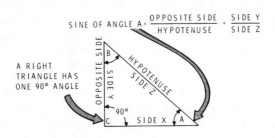

Figure 1-9
The sine function.

The sine of angle A (or angle B) will vary as the angle varies. This is because the length of the opposite side and the length of the hypotenuse must vary accordingly. Angle A or angle B can have any value between 0° and 90°. These are the extreme limits which can occur within the right triangle. Also, the sum of angles A, B, and C must always equal 180°.

As angle A (or B) varies from a minimum of 0° to a maximum of 90°, its sine value will vary from 0 (when side Y is infinitely small) to 1 (when side Y and side Z are equal). The various sine values for all of the angles between 0° and 90° (in increments of 1°) are shown in Figure 1-10. This table of sine values is just a portion of the standard table of trigonometric functions which can be found in almost any textbook on trigonometry.

Angle	Sine	Angle	Sine	Angle	Sine
0°	0.000	31°	.515	61°	.875
1°	.018	32°	.530	62°	.883
2°	.035	33°	.545	63°	.891
3°	.052	34°	.559	64°	.899
4°	.070	35°	.574	65°	.906
5°	.087				
6°	.105	36°	.588	66°	.914
7°	.122	37°	.602	67°	.921
8°	.139	38°	.616.	68°	.927
9°	.156	39°	.629	69°	.934
10°	.174	40°	.643	70°	.940
11°	.191	41°	.656	71°	.946
12°	.208	42°	.669	72°	.951
13°	.225	43°	.682	73°	.956
14°	.242	44°	.695	74°	.961
15°	.259	45°	.707	75°	.966
16°	.276	46°	.719	76°	.970
17°	.292	48°	.731	77°	.974
18°	.309	48°	.743	78°	.978
19°	.326	49°	.755	79°	.982
20°	.342	50°	.766	80°	.985
21°	.358	51°	.777	81°	.988
22°	.375	52°	.788	82°	.990
23°	.391	53°	.799	83°	.993
24°	.407	54°	.809	84°	.995
25°	.423	55°	.819	85°	.996
26°	.438	56°	.829	86°	.998
27°	.454	57°	.830	87°	.999
28°	.470	58°	.848	88°	.999
29°	.485	59°	.857	89°	1.000
30°	.500	60°	.866	90°	1.000

Figure 1-10
A table of sine values for angles
between 0° and 90°.

The value of a sinusoidal waveform increases from zero to maximum in the same manner that the sine values increase as shown in Figure 1-10. This relationship is shown in Figure 1-11. This figure shows a sinusoidal waveform just like the one shown in Figure 1-8. This sine wave is used to represent an ac voltage although it could also be used to represent an ac current. Notice that specific voltage values are observed at 15° intervals between 0° and 90° on the horizontal base line. You must keep in mind that this horizontal line represents the number of degrees of armature rotation. Also the voltage sine wave is assumed to have a maximum value of 1. It could be 1 volt or 1 kilovolt since the actual value is not important. We are concerned only with the relative voltage changes that occur.

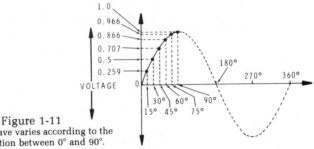

Figure 1-11
The sine wave varies according to the
sine function between 0° and 90°.

As shown in Figure 1-11, the voltage is zero at the beginning of the waveform when the armature is at 0°. When the armature has rotated 15°, the voltage will increase to a value of 0.259 or slightly better than one-fourth of its maximum value. At 30° the voltage has a value of 0.5 or one-half of its maximum value. At 45° the voltage has a value of 0.707 or almost three-fourths of its maximum value. At 60° and 75° the voltage increases to values of 0.866 and 0.966. Then at 90° it reaches its maximum value of 1. If you compare each angular position of the armature (0°, 15°, 30°, etc.) with its corresponding voltage value (0, 0.259, 0.5, etc.) you will find that the voltage increases according to the sine of the angle of rotation. In other words the voltage increases according to the sine values in the table shown in Figure 1-10. If we had examined a greater number of intermediate voltage values between 0° and 90°, we would find that each value would be related to the sine function.

Between the 90 and 180 degree points on the waveform, the voltage drops from its maximum value to zero. During this portion of the waveform, the voltage decreases in the exact opposite manner that it increased. Between the 180 and 360 degree points, the voltage varies in the same manner as it did between 0° and 180°. The only difference is that the polarity of the voltage is opposite. This is why the curve extends below the horizontal line.

The Cycle

Each time the armature of the basic ac generator rotates through one complete revolution (360°), it generates an output voltage that increases and decreases in value in first one direction (or polarity) and then in the opposite direction as shown in Figure 1-12. When the armature rotates one complete revolution, it completes one cycle of events. In other words, it produces one complete change in voltage values. If the armature rotates another 360°, it will simply repeat the cycle or sequence of output voltages. The output voltage produced during one complete revolution of the armature is therefore referred to as one *cycle* of output voltage as indicated in Figure 1-12. In a similar manner the armature will also produce one cycle of output current providing there is a complete circuit through which current can flow.

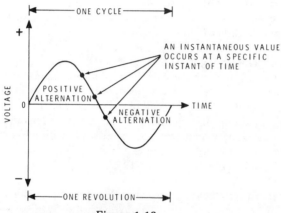

Figure 1-12
One revolution of the armature produces one cycle of output voltage.

During the generation of one cycle of output voltage, there are two changes or alternations in the polarity of the voltage. In other words, the voltage has one polarity during the first half of the cycle and the opposite polarity during the second half. These equal but opposite halves of a complete cycle are referred to as *alternations*. To further distinguish one-half (one alternation) of the cycle from the other, we generally call one alternation the *positive alternation* and the other alternation the *negative alternation*. These two alternations are identified in Figure 1-12. The terms positive and negative are used arbitrarily to distinguish one alternation from the other. During the positive alternation the output voltage will be of a specific polarity and during the negative alternation the polarity of the voltage is reversed.

The voltage sine wave in Figure 1-12 is made up of an infinite number of voltage values which have been plotted and joined together. The voltage at any particular point on the waveform is generally referred to as an *instantaneous* value. An instantaneous value may also be defined as a value which occurs at a specific instant of time. In other words it takes a specific amount of time to generate the sine wave shown in Figure 1-12. For this reason the horizontal line on which the waveform is plotted can be thought of as a time line or a time base and can be calibrated in units of time as well as in degrees of armature rotation. This is why the horizontal line in Figure 1-12 is called a time line.

The horizontal line in Figure 1-12 also serves as a zero reference line. Any voltage value that is plotted on this line will have a value of zero. Above this line all values are assumed to be positive and below the line all values are assumed negative. Therefore, all of the instantaneous voltage values which make up the positive alternation have positive values and all of the instantaneous values which form the negative alternation have negative values as indicated by the plus (+) and minus (−) signs on the vertical axis of the graph shown in Figure 1-12.

Although this discussion may appear to be overly repetitious, the terms described are extremely important and you must be familiar with them. These terms describe the various characteristics of one cycle of a sine wave. You must understand the exact meaning of these terms since they will be used consistently throughout your study of ac electronics.

It is also important to note that the sine wave shown in Figure 1-12 could represent one complete cycle of output current as well as a cycle of output voltage. When current is represented, the positive and negative alternations represent current that is flowing in first one direction and then the opposite direction.

AC VALUES

Since the value of a sine wave of voltage or current continually changes, it is necessary to be specific when describing the value of the waveform. In other words you cannot simply state that a voltage sine wave has a value of 100 volts without specifying if this is the maximum value of the voltage waveform or some value at a specific point between zero and maximum. There are several ways of expressing the value of a sine wave and you must be familiar with each of them. We will now examine the various ways in which the value of a sine wave can be expressed.

Peak Value

As stated earlier, each alternation of a sine wave is made up of an infinite number of instantaneous values. These values are plotted at various heights above and below the horizontal line to form a continuous waveform. It is the height (or amplitude) of each instantaneous value above or below the line that represents the actual value. The greater the height, the higher the value. The points on the waveform which have the greatest height (distance from the horizontal line) are referred to as *peak* values.

The peak values which occur during one cycle of a sine wave of voltage or current are shown in Figure 1-13. Notice that two peak values occur during the cycle. One peak value occurs during the positive alternation when the waveform reaches its maximum height as shown. This point is appropriately identified as the *positive peak value* and it represents the maximum positive value that occurs during the ac cycle.

The second peak value occurs during the negative alternation when the waveform reaches its maximum height below the line. This point is referred to as the *negative peak value* as shown.

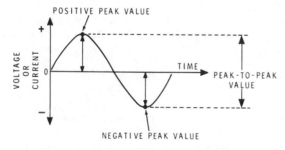

Figure 1-13
Peak and peak-to-peak values.

The positive and negative peak values of a sine wave are equal in value. For example, a voltage sine wave which has a positive peak value of +5 volts would have a negative peak value of −5 volts. Only the polarity of the voltage is different as indicated by the + and − signs. The same relationship is also true for a sine wave of current. The peak value of the current during the positive alternation is equal to the peak value during the negative alternation. The only difference is that the current flows in opposite directions on the positive and negative alternations.

The peak values are sometimes described by using other terms which have the same meaning. For example, the peak value is sometimes called the *peak amplitude* or the *maximum amplitude*. Furthermore, all of these terms apply to any type of waveform. Their use is not limited to the sinusoidal waveform that we are examining at this time.

Peak-To-Peak Value

At times it is necessary to know the total height or value of a sine wave between its peak values. This overall value of the sine wave (from one peak to the other) is called the *peak-to-peak value*. The peak-to-peak value of a sine wave is indicated in Figure 1-13.

The peak-to-peak value can be determined by adding the positive and negative peak values. However, the two peaks are equal in value, thus making it necessary to know only one peak value. In other words the peak-to-peak value can be determined by simply multiplying the one peak value by 2. This relationship is shown mathematically as follows:

$$\text{peak-to-peak value} = 2 \times \text{peak value}$$

For example, if the peak value of a voltage sine wave is equal to 5 volts, the peak-to-peak value must be equal to 2 × 5 or 10 volts. A current sine wave with a peak value of 10 amperes would have a peak-to-peak value that is equal to 2 × 10 or 20 amperes.

It is also possible to determine the peak value, if the peak-to-peak value is known. This can be done by simply dividing the peak-to-peak value by 2 as shown below.

$$\text{peak value} = \frac{\text{peak-to-peak value}}{2}$$

For example, a sine wave with a peak-to-peak value of 18 volts would have a peak value of 18/2 or 9 volts. A current sine wave with a peak-to-peak value of 5 amperes would have a peak value of 5/2 or 2.5 amperes.

It is necessary to understand the meaning and relationship of the peak and peak-to-peak values of a voltage or current sine wave. You will use these values often when analyzing the characteristics of these waveforms and when taking various ac measurements. Various test instruments such as oscilloscopes and certain types of ac voltmeters are used to directly measure the peak-to-peak value of a waveform.

Average Value

When we examine one alternation of a voltage or current sine wave, we find that it increases from zero to a peak value and drops back to zero again as shown in Figure 1-14. The voltage or current remains at its peak value for only an instant and except for this one instant of time, the voltage or current is lower than its peak value. Since the voltage or current remains at the peak value for only an instant, the average voltage for the entire alternation must be less than the peak value. This average voltage can be determined by taking a large number of the instantaneous values which occur during the alternation and computing their average value or it can be performed by using integral calculus. In either case, we would find that the average value of one alternation is equal to 0.636 of its maximum or peak value. This relationship between the peak value and the average value is shown in Figure 1-14 and is summed up in the following equation.

$$\text{average value} = 0.636 \times \text{peak value}$$

For example, a voltage sine wave which has a peak value of 100 volts would have an average value of 0.636 × 100 or 63.6 volts. This same equation may also be used to determine the average value of a current sine wave. For example, a current sine wave with a peak value of 10 amperes would have an average value of 0.636 × 10 or 6.36 amperes.

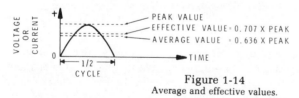

Figure 1-14
Average and effective values.

The equation given above may also be transposed so that the peak value can be determined when the average value is known. When expressed in this manner the equation becomes:

$$\text{peak value} = \frac{\text{average value}}{0.636}$$

For example, suppose the average value of a voltage sine wave is 50 volts. Its peak value would therefore be equal to 50/0.636 or 78.6 volts. A current sine wave, with an average value of 1 ampere, would have a peak value of 1/0.636 or 1.57 amperes.

It is important to note that we have considered the average value of only one alternation or one-half cycle. In order to determine the average value of an entire cycle, we must add the average voltage of one alternation to the average value of the other alternation. Since each alternation has the same average value (0.636 × peak value) and because one value is positive while the other is negative, we must conclude that the two averages cancel to give an overall average of zero. The average value of a full cycle of a sine wave is therefore equal to zero.

Average values are not used extensively when dealing with voltage and current sine waves, but they do have certain special applications. As you proceed with your study of ac electronics you will find that it is beneficial to have a knowledge of how these average values are determined.

Effective Value

When a direct current flows through a resistor, a certain amount of power is dissipated by the resistor in the form of heat. A certain amount of heat is also produced if an alternating current is allowed to flow through the same resistance. However, the heat produced by an alternating current that has a peak value of 1 ampere will not be as great as the heat produced by a direct current that has a value of 1 ampere. The alternating current produces less heat because it reaches its peak value of 1 ampere only once during each alternation and its average value is much lower than 1 ampere. This means that an alternating current with a higher peak value must be used to produce an equivalent amount of heat.

An alternating current that will produce the same amount of heat (in a specified resistance) as a direct current that has a value of 1 ampere, is considered to have an *effective value* of 1 ampere. In other words, a direct current of 1 ampere is equivalent to an alternating current which has an effective value of 1 ampere, as far as their ability to produce heat is concerned. Of course, the alternating current must have a peak value that is higher than 1 ampere in order to be equivalent to the direct current.

34

The effective value of a sine wave of current can be determined by a mathematical process known as the *root-mean-square* or *rms* method. For this reason the effective value is sometimes referred to as an *rms value*. The rms method is lengthy and will not be examined at this time. However, by using this process it can be shown that the effective value of a sine wave of current is equal to 0.707 times its peak value. This relationship between the peak and effective values of a sine wave is shown in Figure 1-14 and is stated below in equation form.

$$\text{effective value} = 0.707 \times \text{peak value}$$

For example, a current sine wave with a peak value of 10 amperes would have an effective value of 0.707 × 10 or 7.07 amperes. This alternating current, which has a peak value of 10 amperes, will produce the same heating effect as a direct current of 7.07 amperes.

Since an alternating current is produced by an alternating voltage, we may also express the ac voltage in terms of its effective value. The effective value of a voltage sine wave is determined by using the same equation given above (0.707 × peak value). For example, a voltage sine wave with a peak value of 40 volts would have an effective value of 0.707 × 40 or 28.28 volts.

This ac voltmeter is used to continuously monitor the 120 volt ac voltage that is used in the home. It is calibrated to read the effective or rms value of an ac sine wave and it can be plugged into any convenient electrical outlet.

The equation just given may be transposed so that the peak value of a sine wave can be determined if its effective value is known. When the equation is appropriately transposed and reduced, we obtain the following equation for the peak value.

$$\text{peak value} = 1.414 \times \text{effective value}$$

This equation may be used to determine the peak value of either a current sine wave or a voltage sine wave. For example, a current sine wave with an effective value of 7 amperes would have a peak value of 1.414×7 or approximately 10 amperes. A voltage sine wave with an effective value of 30 volts would have a peak value of 1.414×30 or 42.4 volts.

Effective (or rms) values are used extensively when working with ac sine waves and it is important that you understand the relationship between the effective and peak values of a sine wave. In most cases, when an ac voltage or current is specified, it is the effective value that is used. In fact, effective values are used so extensively that they are often not specifically identified. For example, it is common practice to express an effective voltage value of 100 volts as simply 100 volts ac or an effective current value of 10 amperes as simply 10 amperes ac. When the alternating current or voltage value is specified, but not specifically identified, the effective value is usually implied. For example, the 120 volt ac electrical power that is used in your home has an effective value of 120 volts. Its peak value is actually much higher than 120 volts as explained earlier.

Most ac voltmeters and ammeters are calibrated to read effective (rms) values although some may be calibrated to indicate peak-to-peak values as well. The ability to accurately measure these effective values is very important, since the effective values are widely used in ac calculations.

Period

When analyzing an ac sine wave, it is often necessary to know just how much time is required to generate one complete cycle of the waveform. The time required to produce one complete cycle is called the *period* of the waveform. The period of a sine wave is shown in Figure 1-15. The period is usually measured in seconds although other units of time can be used. Furthermore, the period is often represented by the letter T as shown.

If a generator produces 1 cycle of output voltage in 1 second, the output sine wave has a period of 1 second. However, if 4 cycles are produced in 1 second, the output sine wave will have a period of ¼th of a second (T = 0.25 seconds). It is important to remember that the period is the time for one cycle only, not the total time required to generate any given number of cycles.

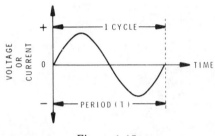

Figure 1-15
The period of a sine wave.

Frequency

It is often necessary to know how rapidly an ac wave form is changing in value. In other words, it may be important to know how many cycles of the waveform occur in a given period of time. The number of cycles that occur in a specified period of time is called the *frequency* of the waveform.

Each time the armature of the simple ac generator (previously described) completes one revolution, one cycle is produced. This means that the frequency of an ac waveform is determined by the speed at which the armature is rotating. As the speed of rotation increases, more cycles are generated in a given period of time, thus causing the frequency to increase.

The frequency of an ac sine wave is usually expressed in terms of the number of cycles generated per second. For example, an armature that rotates 1 complete revolution each second would produce 1 cycle of ac output voltage each second. The ac voltage would therefore have a frequency of 1 cycle per second.

Although the frequency is the number of cycles produced each second, it is expressed in hertz (abbreviated Hz). A generator which produces an ac voltage that completes 1 cycle per second, is said to be operating at a frequency of 1 hertz. The term hertz is simply used in place of cycles per second.

If the ac generator produces 30 cycles of ac output voltage each second, it is operating at a frequency of 30 hertz or 30 Hz. Likewise, an output of 60 cycles each second would be expressed as 60 hertz or 60 Hz.

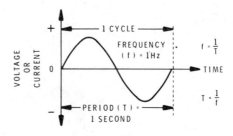

Figure 1-16
The frequency of a sine wave.

There is a definite relationship between the frequency and the period of a sine wave. When the period of a sine wave is equal to 1 second, the frequency will be equal to 1 hertz as shown in Figure 1-16. If the period decreased to 0.5 seconds or one-half of its original value, the frequency would double or increase to 2 hertz. This is because exactly twice as many cycles would occur each second. Likewise, if the period doubled, the frequency would be cut in half.

The relationship between frequency and period is shown in the following equation

$$f = \frac{1}{T}$$

This equation simply states that frequency (represented by the letter f) is equal to 1 divided by the period (T). Furthermore, if the period is expressed in seconds, the frequency obtained will be in hertz (cycles per second). For example, when the period of a sine wave is equal to 0.05 seconds, the frequency of the waveform will be equal to:

$$f = \frac{1}{0.05} = 20 \text{ hertz}$$

A period of 0.05 seconds would therefore correspond to a frequency of 20 hertz. However, if the period was cut in half or reduced to 0.025 seconds, the frequency would be:

$$f = \frac{1}{0.025} = 40 \text{ hertz}$$

In other words the frequency would increase to 40 hertz or double. The equation therefore shows that f and T are inversely proportional. When one increases, the other decreases by a proportional amount and vice versa.

The equation just given may also be transposed so that T can be determined when f is known. This transposed equation is shown below.

$$T = \frac{1}{f}$$

The equation now states that T is equal to 1 divided by f. If f is expressed in hertz, the value of T obtained will be in seconds. For example, when f is equal to 100 hertz, T can be determined as follows:

$$T = \frac{1}{100} = 0.01 \text{ seconds}$$

The period (T) is therefore equal to 0.01 seconds.

Frequencies that range from just a few hertz to many millions of hertz are widely used in the electronics industry. For example, the 120 volt ac electrical power that you use in your home actually has a frequency of 60 hertz. This 60 hertz power is used to operate your lights and appliances and it may even provide heat. In many electronic applications, the frequencies involved are much higher. This is because high frequencies are required to carry information or intelligence. Also, the higher frequencies are easier to convert into electromagnetic (radio) waves which can be transmitted over long distances. These higher frequencies cannot be produced by ac generators since these devices cannot rotate at the very high speeds which would be necessary. Instead, they are produced by electronic circuits which do not require moving parts.

When working with frequencies that extend up to many millions of hertz, you must work with very large numbers which can sometimes be difficult to handle. However, these large numbers can be reduced to a manageable size by using various metric prefixes. The metric prefixes most commonly used for this purpose are defined in Table I below.

Table I

Metric Prefixes		
Prefix	Symbol	Value
kilo	k	1000 (10^3)
mega	M	1,000,000 (10^6)
giga	G	1,000,000,000 (10^9)

These prefixes can be placed before a word to change its meaning. For example, the prefix, *kilo* means 1000 and when it is placed before the word hertz, we obtain the word kilohertz which means 1000 hertz. Generally we express 1000 hertz as simply 1 kilohertz or we use the symbol k to represent kilo and the symbol Hz to represent hertz and express the quantity as 1 kHz. In a similar manner we use mega (M) to represent 1,000,000. Therefore, 1,000,000 hertz can be expressed as 1 megahertz or 1 MHz. The prefix giga (G) represents 1,000,000,000. Therefore, 1,000,000,000 hertz can be expressed as 1 gigahertz or 1 GHz. Also, a frequency of 10,000 hertz could be expressed as 10 kilohertz or a frequency of 1,000,000,000 hertz could be expressed as 1000 megahertz.

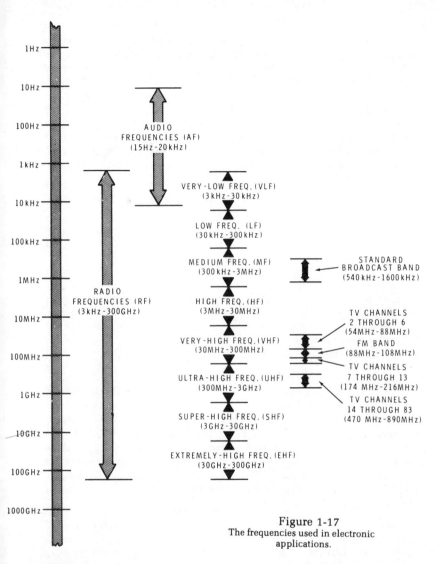

Figure 1-17
The frequencies used in electronic
applications.

The frequencies that are most commonly used in electronic applications are shown in Figure 1-17. As shown in this figure, the frequencies between 15 hertz and 300 GHz (300,000,000,000 hertz) are the ones that are most widely used. The frequencies between 15 hertz and 20 kHz (20,000 hertz) are referred to as audio frequencies (AF) as shown. An ac current that has a frequency within this range will produce an audible tone to which the human ear will respond. However, the ac current must be applied to a device (such as a loudspeaker) which will convert it into sound waves that can be detected by the human ear.

The frequencies between 3 kHz (3000 hertz) and 300 GHz are referred to as radio frequencies (abbreviated RF) since they are used extensively in radio communications. This band of frequencies is further divided into eight smaller bands. The band with the lowest frequencies is called the very low frequency (VLF) band and each higher band is appropriately named. The highest band is called the extremely high frequency (EHF) band. The frequencies within these various RF bands are widely used to transmit information or intelligence from one location to another. Once these frequencies are produced in the form of electrical currents and voltages, they are converted to radio waves which are suitable for transmission through space.

The frequencies within the eight RF bands are further divided into smaller bands which are allocated for specific applications. As shown in Figure 1-17, a small band of frequencies within the medium frequency (MF) band is identified as the standard broadcast band. This band of frequencies is used by commercial AM radio stations. Also notice that the television channels (2 through 6 and 7 through 13) and the FM band (used by FM radio stations) are within the very-high frequency (VHF) band. The higher television channels (14 through 83) are transmitted at higher frequencies which fall within the ultra-high frequency (UHF) band. These various frequency assignments (as well as others which have not been discussed) are controlled by an agency of the U.S. government known as the Federal Communications Commission (FCC).

It is possible to produce frequencies which are higher or lower than those shown in Figure 1-17, but such frequencies are not applicable to most electronic applications. Above the upper RF limit (300 GHz), the characteristics of ac signals change considerably and they are no longer suitable for transmission through a wire or in the form of radio waves. Above this upper limit, we encounter other forms of electromagnetic energy such as light waves, x-rays, and cosmic rays. These various forms of energy are assumed to occur as specified frequencies even though they have many additional characteristics which cannot be explained by conventional electronic theory.

Throughout this discussion on frequency, we have used several metric prefixes to simplify the handling of large numbers. However, additional prefixes are also used to handle extremely small numbers. If you have not studied the various metric prefixes in a previous course or if you feel that you need a comprehensive review, you should read Appendix A at this time. This appendix is a programmed instruction sequence which teaches scientific notation. It shows how various numbers (both large and small) can be expressed as a power of ten and how the appropriate prefixes are assigned to these numbers. It also shows how the various prefixes are used in conjunction with various units of electrical measurement. Be sure to read this appendix carefully if you are not absolutely confident that you understand how to use these prefixes.

NONSINUSOIDAL WAVEFORMS

Although the sine wave is the most basic and widely used ac waveform, it is not the only type of waveform that is used in electronics. In fact, many different types of ac waveforms are used and these waveforms may have very simple or extremely complex shapes.

We will now briefly examine just a few of the more basic ac waveforms that are suitable for a variety of applications. These nonsinusoidal waveforms cannot be produced by the simple ac generator previously described. They are usually generated by electronic circuits which use various types of semiconductor devices such as diodes and transistors.

The Square Wave

Figure 1-18 shows four different types of nonsinusoidal waveforms which could represent either current or voltage. In each case only one cycle of the waveform is shown. The waveform shown in Figure 1-18A is commonly referred to as a *square wave* because its positive and negative alternations are square in shape. The square shape of each alternation indicates that the voltage (or current) immediately rises to its maximum or peak value at one polarity (or direction) and remains there throughout one alternation. Then the voltage immediately changes in polarity (or the current changes direction) and jumps to a peak value and remains there for the duration of the next alternation. When a continuous train of these square waves (one cycle after another) is produced, the voltage or current simply continues to fluctuate back and forth between its peak values.

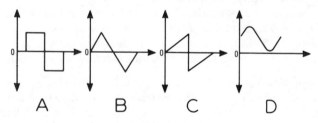

Figure 1-18
Nonsinusoidal waveforms.

43

Not all square waves are symmetrical (with equal positive and negative halves) as shown in Figure 1-18A. In some cases the positive half may be wider (of longer duration) than the negative half or vice versa. Also, some square waves may have a positive peak value that is higher than its negative peak value or vice versa.

Although electrical power can be generated as square waves (by special electronic circuits), the square wave is more useful as an electronic signal since its characteristics can be easily varied.

The Triangular Wave

The waveform shown in Figure 1-18B is called a *triangular wave* because its positive and negative alternations are triangular in shape. Notice that during the positive alternation the waveform rises at a linear rate from zero to a peak value and then decreases back to zero in just the opposite manner. Then on the negative alternation its polarity (or direction) reverses and it again varies in the same manner.

Triangular waves may have peak values that are higher or lower than those shown in Figure 1-18B. In other words, the positive and negative alternations may not always form a perfect triangle which has three equal sides. Triangular waves are used as electronic signals and are seldom used to provide electrical power.

The Sawtooth Wave

The waveform shown in Figure 1-18C is called a *sawtooth wave*. The sawtooth wave is similar to the triangular wave but there are important differences. The sawtooth wave is formed when a voltage or current increases from zero to its positive peak value at a linear rate and then rapidly changes to its negative peak value and then decreases back to zero at a linear rate. This sequence of events represents one complete cycle of the waveform. When a number of cycles are graphically plotted, the waveform has a sawtooth appearance, thus accounting for its name.

The sawtooth waveform may vary slightly from the shape shown in Figure 1-18C. For example, the change from the positive peak value to the negative peak value may not occur almost instantaneously as indicated. This change may require a small, but discernable, amount of time, thus allowing the waveform to more closely resemble a triangular wave.

The sawtooth waveform is often used to control or trigger the operation of electronic circuits. It is used in television sets and in various types of test instruments such as oscilloscopes and digital voltmeters.

Fluctuating DC Waves

There are many instances when an electronic signal does not change direction although it does vary or fluctuate at a specific rate. Such waveforms cannot be truly called ac waveforms, but they still behave as if they were ac signals. For example, the waveform in Figure 1-18D actually represents a dc voltage or current which fluctuates in value according to the sine function. The dc voltage simply fluctuates or rides above the horizontal line which serves as a time base and a zero reference line. This fluctuating dc waveform can produce the same effect as an ac sine wave in certain applications. Such a waveform might appear at an intermediate point within an electronic circuit, but it may be converted into a true ac signal before it reaches the output of the circuit. The waveforms shown in Figure 1-18A, B, and C might also appear as fluctuating dc voltages within the various circuits that generate them and then they might be converted to ac at the appropriate time if it is necessary to do so. In many applications, either an ac or a fluctuating dc signal will work equally well, since it is the manner in which they change that is important, not their specific voltage or current values.

This function generator produces ac sine waves, square waves, and triangular waves over a frequency range of 0.1 hertz to 1 megahertz. The waveforms produced by this instrument are used to test the operation of electronic circuits.

SUMMARY

Unlike direct current (dc), which flows in only one direction and has a steady value, an alternating current (ac) flows in one direction and then the other. Furthermore, the value or magnitude of an alternating current usually varies as it flows in each direction.

Alternating current is used more extensively than direct current because it is more versatile. Alternating current may be used as a source of electrical power or energy or it can be used to carry information or intelligence and thus serve as an electronic signal.

The electrical power that is produced for use in the home is actually alternating current. This ac power is produced by large ac generators at the power plant and distributed through a network of transmission lines to various homes and industries.

Alternating current is used instead of direct current in applications where large amounts of electrical power are required because it is easier and cheaper to produce. Furthermore, an ac voltage can be converted to a higher or lower voltage very easily. AC may also be easily converted into dc.

An alternating current may be used as an electronic signal to carry information from one point to another because its characteristics can be varied in a desired manner. The ac signal may be carried by wires or transmission lines or it may be converted to electromagnetic waves which can be transmitted through space.

An ac generator is able to produce an alternating current because it makes use of a process known as electromagnetic induction. Electromagnetic induction is simply the process of inducing a voltage in a conductor. The induced voltage appears when the conductor moves through a magnetic field.

The voltage induced within a conductor is affected by the strength of the magnetic field, the speed of conductor movement, the length of the conductor in the field, and angle at which the conductor cuts the field. When all of these factors are considered, one simple rule can be formed which states that the voltage induced in a conductor is directly proportional to the rate at which the conductor cuts the magnetic field.

The polarity of the induced voltage is determined by the direction of conductor motion and the direction of the magnetic field.

A simple ac generator is formed by bending a wire into a loop (called an armature) and rotating the armature within a magnetic field. Then slip rings and brushes are used to provide a means of extracting the ac voltage that is induced in the armature.

The ac output voltage produced by the ac generator, varies from zero to maximum and back to zero again as the armature completes one-half of a revolution. Then during the next one-half revolution, the voltage varies in the same manner but its polarity is opposite. One complete revolution of the armature therefore produces an ac voltage that changes in value and polarity. If a load resistance is connected to the generator, the ac output voltage will cause a corresponding ac current to flow through the load.

When the output voltage values are graphically plotted with respect to armature position or time, a curve is formed which shows how the ac voltage varies. Such a curve is called a waveform.

The ac voltage produced by the simple generator varies from zero to its maximum value according to the sine function and then decreases to zero in the exact opposite manner. The ac voltage changes in this manner in each direction. For this reason the ac voltage is said to vary in a sinusoidal manner. When this ac waveform is graphically plotted, it is called a sinusoidal waveform or simply a sine wave.

One complete revolution of the armature produces one cycle of output voltage. One cycle is further divided into two alternations. One alternation is called the positive alternation and corresponds to the first one-half revolution of the armature. The other alternation is called the negative alternation and it corresponds to the second one-half revolution of the armature.

The maximum value reached during each alternation is called a peak value. The total value of a waveform, between its peak values, is called its peak-to-peak value.

The average value of one alternation is equal to 0.636 times the peak value. However, the average value of one complete cycle is zero.

The effective value of an ac sine wave is equal to 0.707 times its peak value. A current sine wave with an effective value of 1 ampere will produce the same amount of heat in a given resistance as a direct current of 1 ampere. The effective value is also called the root-mean-square or rms value.

The period of a sine wave is the time required to produce one complete cycle. The number of cycles that occur in a specified period of time is called the frequency. The period is usually measured in seconds and the frequency is measured in hertz (cycles per second).

A variety of nonsinusoidal waveforms are also used in electronic applications. These waveforms are usually named for their shape. Such waveforms include the square wave, triangular wave, and sawtooth wave.

47

Appendix A

SCIENTIFIC NOTATION

In electronics, it is common to deal with both very large and very small numbers. An example of a very large number is the speed at which electricity travels. It travels at the speed of light which is approximately 1,000,000,000 feet per second or about 300,000,000 meters per second. As for very small numbers, consider the size and weight of an electron. It is believed that the electron has a diameter of approximately 0.000 000 000 0022 inch and a weight of about 0.000 000 000 000 000 000 000 000 0009 gram. Sometimes, we perform arithmetic with numbers such as these. To simplify such arithmetic, a shorthand method has been developed to express numbers. This shorthand method is called *scientific notation*. The following programmed instruction sequence will serve as an introduction to scientific notation.

---

1. As mentioned above, scientific notation is a shorthand method of expressing numbers. While any number can be expressed in scientific notation, this technique is particularly helpful in expressing very large and very _____ numbers.

---

2. (small) Scientific notation is based on a concept called powers of ten. Thus, in order to understand scientific notation we should first learn what is meant by powers of _____.

---

3. (ten) In mathematics, a number is raised to a power by multiplying the number times itself one or more times. Thus, we raise 5 to the second power by multiplying 5 times itself. That is, 5 to the second power is $5 \times 5 =$_____.

---

4. (25) Also, 5 to the third power is the same as saying $5 \times 5 \times 5 =$ _____.

5. (125) Thus, 5 can be raised to any power simply by multiplying it times itself the required number of times. For example, $5 \times 5 \times 5 \times 5 = 625$. Consequently, 5 raised to the _____ power is equal to 625.

6. (fourth) The above examples use powers of five. However, any number can be raised to a power by the technique of multiplying it times itself the required number of times. Thus, the powers of two would look like this:

> 2 to the second power equals $2 \times 2 = 4$
>
> 2 to the third power equals $2 \times 2 \times 2 = 8$
>
> 2 to the fourth power equals $2 \times 2 \times 2 \times 2 = 16$
>
> 2 to the fifth power equals $2 \times 2 \times 2 \times 2 \times 2 = 32$
>
> 2 to the sixth power equals $2 \times 2 \times 2 \times 2 \times 2 \times 2 =$ _____.

7. (64) In mathematics, the number which is raised to a power is called the *base*. If 5 is raised to the third power, 5 is considered the

_____.

8. (base) The power to which the number is raised is called the *exponent*. If 5 is raised to the third power, then the exponent is 3. In the same way, if 2 is raised to the sixth power, then 2 is the base while 6 is the _____.

9. (exponent) There is a shorthand method for writing "2 raised to the sixth power." It is:

$$2^6$$

Notice that the exponent is written as a small number at the top right of the base. Remember _this_ number is the base while _this_ number is the exponent.

$$2^6$$

Therefore in the example 3^4, 3 is the _____ while 4 is the

_____.

10. (base, exponent) The number 3^4 is read "3 raised to the fourth power." It is equal to:

$$3 \times 3 \times 3 \times 3 = 81.$$

The number 4^6 is read _____.

---

11. (4 raised to the sixth power) Scientific notation uses powers of ten. Several powers of ten are listed below:

$$10^2 = 10 \times 10 = 100$$
$$10^3 = 10 \times 10 \times 10 = 1000$$
$$10^4 = 10 \times 10 \times 10 \times 10 = 10,000$$
$$10^5 = 10 \times 10 \times 10 \times 10 \times 10 = 100,000$$
$$10^6 = 10 \times 10 \times 10 \times 10 \times 10 \times 10 = \text{_____}.$$

---

12. (1,000,000) Multiplication by 10 is extremely easy since all we have to do is add one zero for each multiplication. Another way to look at it is that multiplication by ten is the same as moving the decimal point one place to the right. Thus, we can find the equivalent of 10^2 by multiplying $10 \times 10 = 100$; or, simply by adding a 0 after 10 to form 100; or by moving the decimal point one place to the right to form $10.0 = 100$. In any event, 10^2 is equal to _____.

---

13. (100) There is a simple procedure for converting a number expressed as a power of ten to its equivalent number. We simply write down a 1 and after it write the number of zeros indicated by the exponent. For example, 10^6 is equal to 1 with 6 zeros after it. In the same way 10^{11} is equal to 1 with _____ zeros after it.

---

14. (11) This illustrates one of the advantages of power of ten. It is easier to write and remember 10^{21} than its equivalent number: 1,000,000,000,000,000,000,000. Try it yourself and see if it isn't easier to write 10^{35} than to write its equivalent number of: _____.

15. (100,000,000,000,000,000,000,000,000,000,000). In the previous examples, we converted a number expressed in powers of ten to its equivalent number. Now let's see how we convert in the opposite direction. Remember the number must be expressed using 10 as the base with the appropriate exponent. The exponent is determined simply by counting the zeros which fall on the right side of the 1. Thus, 1,000,000 becomes 10^6 because there are 6 zeros in the number. In the same way, 10,000,000,000 is expressed as _____.

16. (10^{10}) To be sure you have the right idea, study each of the groups below. Which group contains an error? _____.

Group A	Group B	Group C
$10^6 = 1,000,000$	$1000 = 10^3$	$10^7 = 10,000,000$
$10^2 = 100$	$10,000 = 10^4$	$10^9 = 1,000,000,000$
$10^9 = 1,000,000,000$	$100 = 10^2$	$10^{11} = 10,000,000,000$

17. (Group C) There are two special cases of powers of ten which require some additional explanation. The first is 10^1. Here the exponent of 10 is 1. If we follow the procedure developed in Frame 13 we find that $10^1 = 10$. That is, we put down a 1 and add the number of zeros indicated by the exponent. Thus $10^1 =$ _____.

18. (10) The other special case is 10^0. Here the exponent is 0. Once again we follow the procedure outlined in Frame 13. Here again we write down a 1 and add the number of zeros indicated by the exponent. However, since the exponent is 0, we add no zeros. Thus, the equivalent number of 10^0 is 1. That is $10^0 =$ _____.

19. (1) Any base number with an exponent of 1 is equal to the base number. Any base number with an exponent of 0 is equal to 1. Thus, $X^1 =$ _____ and $X^0 =$ _____.

20. (X, 1) In the examples given previously, the exponents have been positive numbers. For simplicity the plus sign has been omitted. Therefore, 10^2 is the same as 10^{+2}. Also, 10^6 is the same as _____.

21. (10^{+6}) Positive exponents represent numbers larger than 1. Thus, numbers such as 10^2, 10^7, and 10^{15} are greater than 1 and require _____ exponents.

22. (positive) Numbers smaller than 1 are indicated by negative exponents. Thus, numbers like 0.01, 0.0001, and 0.00001 are expressed as negative powers of ten because these numbers are less than _____.

23. (1) Some of the negative powers of ten are listed below:

$10^{-1} = 0.1$
$10^{-2} = 0.01$
$10^{-3} = 0.001$
$10^{-4} = 0.0001$
$10^{-5} =$ _____.

24. (0.00001) A brief study of this list will show that this is simply a continuation of the list shown earlier in Frame 11. If the two lists are combined in a descending order, the result will look like this:

$10^6 = 1,000,000.$
$10^5 = 100,000.$
$10^4 = 10,000.$
$10^3 = 1,000.$
$10^2 = 100.$
$10^1 = 10.$
$10^0 = 1.$
$10^{-1} = 0.1$
$10^{-2} = 0.01$
$10^{-3} = 0.001$
$10^{-4} = 0.0001$
$10^{-5} =$ _____.

25. (0.00001) We can think of the negative exponent as an indication of how far the decimal point should be moved to the left to obtain the equivalent number. Thus, the procedure for converting a negative power of ten to its equivalent number can be developed. The procedure is to write down the number 1 and move the decimal point to the left the number of places indicated by the negative exponent. For example, 10^{-4} becomes:

$$0.0001. \text{ or } 0.0001$$

Notice that the -4 exponent indicates that the decimal point should be moved _____ places to the _____.

26. (4, left) Up to now we have used powers of ten to express only those numbers which are exact multiples of ten such as 100, 1000, 10,000, etc. Obviously, if these were the only numbers which could be expressed as powers of ten, this method of writing numbers would be of little use. Actually, any _____ can be expressed in powers-of-ten notation.

27. (number) The technique by which this is done can be shown by an example. If 1,000,000 can be represented by 10^6, then 2,000,000 can be represented by 2×10^6. That is, we express the quantity as a number multiplied by the appropriate power of ten. As another example, $2,500,000 = 2.5 \times 10^6$. Also, $3,000,000 =$ _____.

28. (3×10^6) In the same way, we can write 5,000 as 5×10^3. Some other examples are:

$$200 = 2 \times 10^2$$
$$1500 = 15 \times 10^2$$
$$22,000 = 22 \times 10^3$$
$$120,000 = 12 \times 10^4$$
$$1,700,000 = 17 \times 10^5$$
$$9,000,000 = \underline{\hspace{2cm}} .$$

29. (9×10^6) By the same token, we can convert in the opposite direction. Thus, 2×10^5 becomes $2 \times 100,000$ or $200,000$. Also, $2.2 \times 10^3 = 2.2 \times 1000 = 2,200$. And, $66 \times 10^4 = $ _____.

30. ($660,000$) You may have noticed that when we use powers of ten there are several different ways to write a number. For example, $25,000$ can be written as 25×10^3 because 25×1000 equals $25,000$. However; it can also be written as 2.5×10^4 because $2.5 \times 10,000$ equals $25,000$. It can even be written as 250×10^2 since $250 \times 100 = 25,000$. In the same way, 4.7×10^4, 47×10^3, and 470×10^2 are three different ways of writing the number _____.

31. ($47,000$) Numbers smaller than one are expressed as negative powers of ten in much the same way. Thus, $.0039$ can be expressed as 3.9×10^{-3}, 39×10^{-4}, or $.39 \times 10^{-2}$. Also, 6.8×10^{-5}, 68×10^{-6}, and $.68 \times 10^{-4}$ are three different ways of expressing the number _____.

32. ($.000068$) As you can see there are several different ways in which a number can be written as a power of ten. Scientific notation is a way of using powers of ten so that all numbers can be expressed in a uniform way. To see exactly what scientific notation is, consider the following examples of numbers written in scientific notation:

$$6.25 \times 10^{18}$$
$$3.7 \ \times 10^6$$
$$4.0 \ \times 10^2$$
$$6.8 \ \times 10^{-4}$$
$$3.9 \ \times 10^{-6}$$
$$2.2 \ \times 10^{-12}$$

Notice that the numbers range from a very large number to an extremely small number. And yet, all these numbers are written in a uniform way. This method of writing numbers is called scientific _____.

33. (notation) The rules for writing a number in scientific notation are quite simple. First, the decimal point is always placed after the first digit on the left which is not a zero. Therefore, the final number will appear in this form: 6.25, 7.3, 9.65, 8.31, 2.0 and so forth. It must never appear in a form such as: .625, 73, 96.5, .831 or 20. Thus, there is always one and only one digit on the _____ side of the decimal point.

34. (left) The second rule involves the sign of the exponent. If the original number is greater than 1, the exponent must be positive. If the number is less than 1, the exponent must be negative. Thus, 67,000 requires a positive exponent but 0.00327 requires a _____ exponent.

35. (negative) Finally, the magnitude of the exponent is determined by the number of places that the decimal point is moved. For example, 39,000.0 is expressed as 3.9×10^4 because the decimal point must be moved 4 places in order to have only one digit to the left of it. Using this rule, 6,700,000,000 is expressed as $6.7 \times$ _____.

36. (10^9) The number 0.00327 is expressed as 3.27×10^{-3}. Here the decimal point is moved 3 places in order to have one digit which is not zero to the left of the decimal. Likewise 0.00027 is expressed as $2.7 \times$ _____.

37. (10^{-4}) To be sure you have the idea look at the groups of numbers below. Which of the following groups contains a number that is not expressed properly in scientific notation? _____

Group A	Group B	Group C
6.25×10^{18}	1.11×10^{11}	6.9×10^{10}
3.75×10^{-9}	$-3.1 \ \times 10^2$	3.4×10^7
4.20×10^1	$-3.1 \ \times 10^{-2}$	39.5×10^2
7.93×10^0	2.00×10^2	6.0×10^4

38. (Group C) The number 39.5×10^2 is not written in scientific notation because there are two digits on the left side of the decimal point. The minus signs in Group B may have confused you. Although, it has not been mentioned, negative numbers can also be expressed in scientific notation. Thus, a number like $-6,200,000$ becomes -6.2×10^6. All the rules previously stated hold true except that now a _____ sign is placed before the number.

39. (minus) Small negative numbers are handled in the same way. Thus -0.0092 becomes -9.2×10^{-3}. The minus sign before the number indicates that this is a negative number. The minus sign before the exponent indicates that this number is less than _____.

40. (1) Listed below are numbers which are converted to scientific notation. Which one of these groups contains an error? _____

Group A	Group B	Group C
$2,200 = 2.2 \times 10^3$	$119,000 = 1.19 \times 10^5$	$119 = 1.19 \times 10^2$
$32,000 = 3.2 \times 10^4$	$1,633,000 = 1.633 \times 10^6$	$93 = 9.3 \times 10^1$
$963,000 = 9.63 \times 10^5$	$937,000 = 9.37 \times 10^4$	$7.7 = 7.7 \times 10^0$
$660 = 6.6 \times 10^2$	$6,800 = 6.8 \times 10^3$	$131.2 = 1.312 \times 10^2$

41. (Group B) 937,000 converts to 9.37×10^5 and not to 9.37×10^4. Which of the groups below contains an error? _____

Group A	Group B	Group C
$0.00037 = 3.7 \times 10^{-4}$	$0.44 = 4.4 \times 10^{-1}$	$.37 = 3.7 \times 10^{-1}$
$0.312 = 3.12 \times 10^{-1}$	$0.0002 = 2.0 \times 10^{-4}$	$.0098 = 9.8 \times 10^{-3}$
$0.068 = 6.8 \times 10^{-2}$	$0.0798 = 7.98 \times 10^{-2}$	$.00001 = 1.0 \times 10^{-5}$
$0.0092 = 9.2 \times 10^3$	$0.644 = 6.44 \times 10^{-1}$	$0.0075 = 7.5 \times 10^{-3}$

42. (Group A) The final number in group A requires a negative exponent. Which of the groups below contains an error?_____

Group A	Group B	Group C
$3{,}700{,}000 = 3.7 \times 10^6$	$9440 = 9.44 \times 10^3$	$20 = 2.0 \times 10^1$
$-5{,}500 = -5.5 \times 10^3$	$-110 = -1.1 \times 10^2$	$0.02 = 2.0 \times 10^{-2}$
$0.058 = 5.8 \times 10^{-2}$	$0.0062 = 6.2 \times 10^{-4}$	$-200{,}000 = -2.0 \times 10^5$
$-0.0034 = -3.4 \times 10^{-3}$	$-0.0123 = -1.23 \times 10^{-2}$	$-0.000200 = -2.0 \times 10^{-4}$

43. (Group B) 0.0062 is equal to 6.2×10^{-3}. Match the following:

1. 16   a. 1.6×10^{-3}
2. .0016   b. 1.6×10^4
3. 160,000   c. 1.6×10^0
4. 1.6   d. 1.6×10^1
5. .016   e. 1.6×10^{-2}
6. 16,000   f. 1.6×10^5

44. (1-d, 2-a, 3-f, 4-c, 5-e, 6-b) Another concept that goes hand in hand with powers of ten and scientific notation is metric prefixes. These are prefixes such as *mega* and *kilo* which when placed before a word change the meaning of the word. For example, the prefix *kilo* means *thousand*. When kilo and meter are combined the word kilometer is formed. This word means 1000 meters. In the same way, the word kilogram means _____ grams.

45. (1,000) Since kilo means 1,000 we can think of it as multiplying any quantity times 1000 or 10^3. Thus, kilo means 10^3. Another popular metric prefix is *mega*. *Mega* means *million*. Thus a megaton is one million tons or 10^6 tons. In the same way one million volts is referred to as a _____volt.

46. (mega) One thousand watts can be called a kilowatt. Also one million watts can be called a _____.

47. (megawatt) A kilowatt is equal to 10^3 watts while a megawatt is equal to _____ watts.

48. (10^6) Often it is convenient to convert from one prefix to another. For example, since a megaton is 10^6 tons and a kiloton is 10^3 tons, a megaton equals 1000 kilotons. And, since a megaton is one thousand times greater than a kiloton, the kiloton is equal to .001 megaton. Now, consider the quantity 100,000 tons. This is equal to 100 kilotons or _____ megatons.

49. (0.1) Kilo is often abbreviated k. Thus, 100 kilowatts may be expressed as 100 k watt. Mega is abbreviated M. Therefore 10 megawatts may be expressed as _____ watts.

50. (10 M) The quantity 5 k volts is 5 kilovolts or 5000 volts. Also, 5 M volts is 5 megavolts or _____ volts.

51. (5,000,000) There are also prefixes which have values less than one. The most used are:

milli — which means *thousandths* (.001) or 10^{-3}
micro — which means *millionths* (.000 001) or 10^{-6}.

One thousandth of an ampere is called a milliampere. Also, one thousandth of a volt is called a _____ .

52. (millivolt) If a second is divided into one million equal parts each part is called a microsecond. Also, the millionth part of a volt is called a _____.

53. (microvolt) One volt is equal to 1000 millivolts or 1,000,000 microvolts. Or, 1 volt equals 10^3 millivolts and 10^6 microvolts. Expressed another way, 1 millivolt equals .001 volt while 1 microvolt equals .000001 volt. Thus, 1 millivolt equals 10^{-3} volts while 1 microvolt equals _____ volt.

54. (10^{-6}) Powers of ten allow us to express a quantity using whichever metric prefix we prefer. For example, we can express 50 millivolts as 50×10^{-3} volts simply by replacing the prefix milli with its equivalent power of ten. In the same way 50 microvolts is equal to $50 \times$_____ volts.

55. (10^{-6}) When writing abbreviation for the prefix milli the letter small *m* is used. A small *m* is used to distinguish it from mega which used a capital M. Obviously, the abbreviation for micro cannot also be *m*. To represent micro the Greek letter μ (pronounced mu) is used. Thus, 10 millivolts is abbreviated 10 m volts while 10 microvolts is abbreviated 10 μ volts. Remember, m means 10^{-3} while μ means _____.

56. (10^{-6}) Match the following:

1.	M watt	a.	10^{-3} watts
2.	k watt	b.	10^{-6} watts
3.	m watt	c.	500×10^{-3} watts
4.	μ watt	d.	10^6 watts
5.	.5 watt	e.	.5 k watts
6.	500 watts	f.	10^3 watts
7.	500,000 watts	g.	.5 M watts
8.	.00005 watts	h.	.05 k watts
9.	50 watts	i.	5 m watts
10.	.005 watts	j.	50 μ watts

57. (1-d, 2-f, 3-a, 4-b, 5-c, 6-e, 7-g, 8-j, 9-h, 10-i) Additional aspects of powers of ten, scientific notation, and metric prefixes will be discussed later.

Unit 2

AC MEASUREMENTS

INTRODUCTION

Anyone that operates, repairs, or designs electronic equipment must know how to measure the various ac quantities such as current, voltage, and power. These ac quantities are usually measured with specially designed ac meters or test instruments. A variety of electromechanical meters are used for this purpose and most of these meters are inexpensive and easy to operate. Most of these inexpensive meters are designed to measure effective or rms values although some are also calibrated to indicate peak-to-peak values.

Other test instruments, which are more complex and more expensive, are used to measure additional ac characteristics. These more sophisticated instruments can measure instantaneous and peak-to-peak ac values as well as the frequency and period of various types of sinusoidal and nonsinusoidal waveforms.

In this unit you will examine some of the most important test instruments that are used to take various ac measurements. You will learn basically how these test instruments operate and you will see how they are used in typical applications. After you have analyzed these basic instruments, you will examine some simple ac circuits which contain only resistive components.

AC METERS

A variety of ac meters are used to measure alternating current and voltage values. Most of these meters are electromechanical devices which depend on induced magnetism for operation. We will now examine some of the most widely used ac meters. We will see how these meters operate and how they are used.

Rectifier-Type, Moving-Coil Meters

One of the most widely used ac meters utilizes a *moving-coil* meter movement in conjunction with a group of rectifier diodes. The moving-coil meter is actually designed to measure dc current and the rectifier diodes (called rectifiers) are used to convert the ac current into a dc current which will operate the meter.

The Basic Meter Movement. The moving-coil meter movement is also commonly referred to as a d'Arsonval meter movement in honor of its inventor Arsene d'Arsonval. This meter movement is essentially the heart of the rectifier-type, moving-coil meter.

A typical moving-coil meter movement is shown in Figure 2-1. A horseshoe magnet is used to produce a stationary magnetic field which cuts across a moving coil as shown. The moving coil consists of many turns of fine wire on an aluminum frame and the coil is mounted so that it can rotate within the magnetic field much like the armature of an ac generator. However, the coil cannot rotate 360° or one complete revolution. Instead, it can only move between specific limits. This is because a

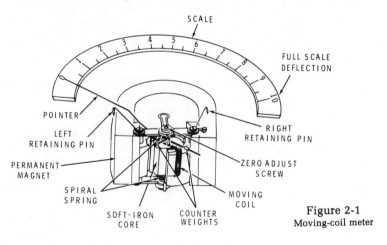

Figure 2-1
Moving-coil meter

62

pointer (or indicator needle) is attached to one end of the coil so that it moves or swings as the coil rotates. This pointer is allowed to move only between the left and right retaining pins as shown. Also, the moving coil rotates around a soft iron core which is fixed in place. This iron core helps to maintain a uniform magnetic field between the opposite poles of the magnet.

The spiral springs shown in Figure 2-1 are used to force the coil and its associated pointer to the extreme left so that the pointer is near the left retaining pin. At this time the pointer rests above the zero calibration mark on the meter's scale. These spiral springs are also used to apply current to the moving coil. The two ends of the moving coil connect to the inner ends of the spiral springs. The outer end of the rear spring is fixed in place but the outer end of the front spring connects to a zero adjust screw. By adjusting this screw, the tension on the spring may be controlled and the pointer may be positioned. Normally, the screw is adjusted so that the pointer will indicate zero when no current is flowing through the moving coil. Counterweights are attached to the pointer so that it will be perfectly balanced and rotate smoothly on its pivots.

The action that takes place in the moving-coil meter movement is illustrated in Figure 2-2. Figure 2-2A shows a front view of the basic meter movement. As shown, the moving coil rotates around the soft-iron core and remains within the magnetic field. This illustration also shows the relative position of the pole pieces which are attached to the opposite poles of the magnet. These pole pieces help to maintain a uniform magnetic field through which the moving coil must rotate.

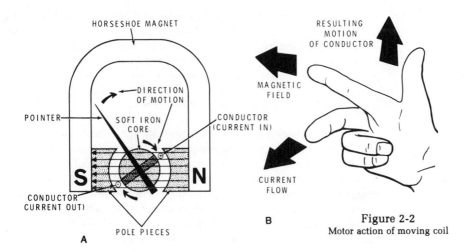

Figure 2-2
Motor action of moving coil

The meter movement in Figure 2-2 has been simplified so that only one turn of the moving coil is shown. This single turn is viewed from one end or in other words, a cross sectional view is shown. When current is forced through the coil so that it flows into the upper right conductor and out of the lower left conductor, a unique action occurs. The current through the conductor creates a magnetic field which surrounds the conductor and this field interferes with the stationary field produced by the magnet. The interaction of the fields causes the coil to move. This action is exactly opposite to the generator action described in the previous unit. In an ac generator, an armature moves through a magnetic field and a voltage (and corresponding current) is induced within the armature loop. However, in this case a current is forced through a loop of wire and the loop is forced to turn because the fields interact. Electrical energy is therefore converted to mechanical energy instead of mechanical energy being converted to electrical energy.

The moving-coil meter movement operates on the same basic principle as an electric motor. The moving-coil responds to current just like the armature in a motor. Although it is possible to analyze the exact manner in which the fields interact to determine the direction of conductor motion, a simple rule can be used to determine the same end result. This rule is commonly referred to as the *right-hand motor rule* and it is illustrated in Figure 2-2B. When the right hand is positioned as shown, with the forefinger pointing in the direction of the magnetic field and the middle finger (which is bent out from the palm at 90°) pointing in the direction of current flow, the thumb points in the direction of conductor motion.

When the right hand rule is applied to the coil in Figure 2-2A, we find that the upper right side of the coil is forced down and the lower left side is forced up. Therefore, the pointer is forced to move up scale from left to right or in a clockwise direction. The coil must move against the tension provided by the spiral springs, and the resulting pointer deflection is proportional to the amount of current flowing through the coil. The higher the current, the greater the deflection.

Like all meter movements, the moving-coil meter movement is rated according to the amount of current required to produce full-scale deflection. For example, a 1 milliampere meter movement would deflect full scale when 1 milliampere of current flows through it. Some meters are designed to be more sensitive than others. For example, a meter with a 200 microampere rating would be more sensitive than one with a 1 milliampere rating. Other commonly used meters which are even more sensitive, have ratings of 100 and 50 microamperes.

It is also important to note that all moving-coil meter movements are designed to accept only direct current. These meters will not respond properly if alternating current is applied directly to them.

The Rectifiers. The moving-coil meter movement may be used to measure alternating current if the ac is converted to dc before it is applied to the meter movement. This is usually accomplished by using a group of rectifier diodes (also called rectifiers). These rectifiers are connected between the input ac and the meter and they allow current to flow in only one direction through the meter.

One of the early rectifier diodes that was once commonly used in conjunction with meter movements is illustrated in Figure 2-3A. It consists of a copper disk which has a layer of copper oxide on one side. A lead disk is placed against the copper oxide layer and the entire unit is compressed between two metal pressure plates which are held in place by a bolt. This type of rectifier diode is commonly referred to as a *copper oxide rectifier*.

The copper oxide rectifier will allow current to flow readily from the copper to the copper oxide, but will allow only a small current to flow in the reverse direction. Therefore, it is essentially a unidirectional conductor of electrical current. An ideal rectifier would act as a short in one direction and an open in the other direction. However, this ideal situation cannot be achieved in practice and most practical rectifiers have substantially less than ideal characteristics.

The copper oxide rectifier has now been replaced in most applications by a more efficient device known as a *junction diode rectifier* or simply a *junction diode*. The junction diode is made from a semiconductor material (a material which is not a good conductor or a good insulator) such as germanium or silicon. The semiconductor material is processed in a unique manner so that it takes on unidirectional characteristics. The semiconductor device is then mounted in a glass, metal, or plastic package which protects it from environmental hazards. A typical junction diode is shown in Figure 2-3B. The device shown has two metal leads which extend from opposite ends of a plastic case. The leads are internally connected to the opposite ends of the junction diode rectifier. Current can readily flow into one lead and out of the other lead in the direction shown but only a small leakage current can flow in the opposite direction.

One end of the rectifier shown in Figure 2-3B is marked with a band. This end of the diode is generally referred to as the *cathode* and the opposite end is called the *anode*. Current always flows from the cathode end to the anode end as shown.

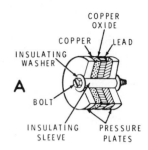

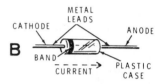

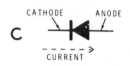

Figure 2-3
A typical copper oxide rectifier (A), junction diode rectifier (B), and rectifier symbol (C).

65

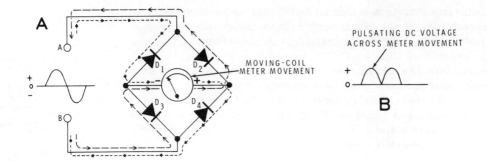

Figure 2-4
A rectifier-type, moving-coil meter

The schematic symbol that is commonly used to represent a rectifier (of any type) is shown in Figure 2-3C. Notice that the symbol consists of a bar and an arrow and that current (electron) flow is always opposite to the direction that the arrow is pointing as shown. In other words, current can flow from the bar to the arrow but not in the opposite direction.

The Complete AC Meter. The schematic diagram of a rectifier-type, moving-coil meter is shown in Figure 2-4A. Notice that four rectifiers are used in conjunction with one meter movement. The four rectifiers are identified as D_1, D_2, D_3, and D_4 and are arranged in what is called a *bridge rectifier* configuration. The two input terminals of the circuit are identified as A and B.

Assume that an ac generator is connected to input terminals A and B and that this generator is supplying a continuous alternating current which varies in a sinusoidal manner. Also assume that during each positive alternation of the ac sine wave, terminal A is positive with respect to terminal B. During each positive alternation the circuit current would therefore be forced to flow along the path indicated by the long dashed lines. It would flow from terminal B, through D_3, through the meter movement, through D_2, and back to terminal A. During each negative alternation, when B is positive with respect to A, current must flow along the path indicated by the short dashed lines. In other words, it would flow from terminal A, through D_1, through the meter movement, through D_4, and back to terminal B. Notice that even though the input current changes direction, the current through the meter movement is always in the same direction. The four rectifiers therefore convert the input ac into dc (actually a pulsating dc) as shown in Figure 2-4B.

66

The process of converting ac to dc is referred to as *rectification* and this is why the diodes used in this application are referred to as simply rectifiers. Furthermore, the rectifiers in this circuit convert both halves of the ac sine wave into a pulsating direct current and are therefore said to be providing *full wave* rectification. Less complicated circuits, which use only one or two rectifiers, may also be used to provide *half wave* rectification. In this last process, only one-half of the ac sine wave is allowed to flow through the meter movement.

The current through the meter movement flows in pulses since each alternation rises from zero to a peak value and drops back to zero again. Unless the frequency of the ac input is extremely low, the meter movement will not be able to follow the variations in the pulsating current. Instead, the meter's pointer responds to the average value of the ac sine wave, or in other words, 0.636 times the peak value. However, the scale on the meter is usually calibrated in effective or rms values. In other words, the numbers on the meter scale represent effective values which are equal to 0.707 times the peak value. The effective value of a sine wave is much more important than the average value since effective values are used in most ac calculations involving current and voltage.

Electrical Characteristics. A variety of rectifier-type, moving-coil meters are available which are capable of measuring a wide range of alternating currents. Each meter is designed to measure up to a certain maximum current. For example, some meters may have calibrated scales which extend from 0 to 1 milliampere, 0 to 10 milliamperes, or 0 to 100 milliamperes.

Most rectifier-type, moving-coil meters have an accuracy of ±5%. This accuracy measurement is based on the percentage of error (in the meter reading) at full-scale deflection. For example, a 100 milliampere meter which has an accuracy of ±5% might be off by as much as ±5 milliamperes when it is indicating a current of 100 milliamperes. However, this same full scale accuracy of ±5% will pertain to any other meter reading. For example, if the meter indicates 50 milliamperes, the true reading could still be off by as much as ±5 milliamperes.

The scale used with the rectifier-type, moving-coil meter is linear. This simply means that the numbers or values on the scale are equally spaced. Therefore, the pointer's deflection is always directly proportional to the current flowing through the meter movement. A typical scale used with this type of meter is shown in Figure 2-5.

Figure 2-5
A typical linear scale

The rectifier-type, moving-coil meter is useful for measuring alternating currents over a specific frequency range. This type of meter is usually quite accurate over a frequency range that extends from approximately 10 hertz to as much as 10,000 to 15,000 hertz. Below a lower limit of approximately 10 hertz, the meter's pointer tends to fluctuate in accordance with the changes in input current, thus making it difficult to read the meter. Above the upper frequency limits just mentioned, the meter readings are usually too inaccurate to be usable. In fact, the accuracy of the meter progressively gets worse as frequency increases beyond just a few hundred hertz because the rectifiers have a certain amount of internal *capacitance* and the moving-coil meter movement has a certain amount of internal *inductance*. These two internal properties (to be discussed in later units) present a certain amount of opposition to the flow of alternating current through the meter and this opposition varies with frequency.

Moving-Vane Meters

The meter previously described requires the use of rectifiers to convert the ac input into dc which is needed to operate the meter movement. However, some meters are capable of responding directly to the ac input. One of these devices is commonly referred to as a *moving-vane* meter or a *moving-iron* meter. This type of meter contains a movable iron vane and depending on the shape of the vane, it may be classified as either a *radial-vane* or a *concentric-vane* meter. We will now examine both of these basic types and then consider their important electrical characteristics.

The Radial-Vane Meter Movement. The basic radial-vane mechanism is shown in Figure 2-6. Notice that it contains essentially the same basic components as the meter movement previously described. However, these components are arranged in basically the opposite manner. In other words, a coil is used in this meter movement but this coil is stationary and it surrounds a moving iron vane which is attached to the meter's pointer as shown. Also, a stationary iron vane is mounted inside of the coil so that

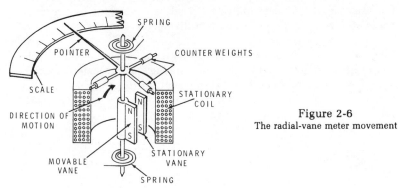

Figure 2-6
The radial-vane meter movement

68

it is lined up parallel to the moving vane. When current flows through the coil (in either direction) a magnetic field is produced which surrounds the coil. This field passes through the moving and stationary vanes in the same direction and magnetizes the two vanes in the same direction. Therefore, the vanes will always have north and south poles which are directly adjacent to each other. Since a fundamental rule of magnetism states that like poles must repel each other, the movable vane is pushed away from the stationary vane and the meter's pointer is forced to rotate against the tension provided by the springs. Figure 2-6 shows the tops of the vanes to be north poles while the bottoms are south poles. This condition will occur only when current flows in one specific direction. When current flows in the opposite direction, the poles are reversed. However, in either case, the two vanes are pushed apart.

A higher current in the coil will produce a stronger magnetic field around the coil which results in greater induced magnetism within the vanes which in turn causes the pointer to deflect further. In other words, the higher the current the greater the pointer deflection. As shown, the scale must be appropriately calibrated so that the pointer will indicate the amount of current applied to the meter movement. Like most ac meters, the scale is usually calibrated in effective values and extends from zero to the maximum value that the meter is designed to measure.

The Concentric-Vane Meter Movement. The basic concentric-vane meter movement is shown in Figure 2-7. Notice that it is similar to the radial-vane instrument but its stationary and moving iron vanes are semicircular in shape. The moving vane is mounted inside of the stationary vane and it is attached to the pointer. The moving vane has square edges but the stationary vane is tapered along one edge.

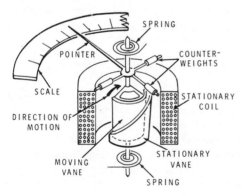

Figure 2-7
The concentric-vane meter movement

When current flows through the coil, the magnetic field produced around the coil passes through the two vanes and causes them to become magnetized in the same direction. However, the magnetic lines of force are not distributed uniformly through the stationary vane because of its tapered edge. Fewer lines of force will pass through the narrow end than through the wide end because the narrow end offers higher resistance or opposition to the magnetic lines of force. The wider end provides an easier path for the lines of force, thus allowing more lines of force to be produced. Therefore, the wider end of the vane becomes more strongly magnetized than the narrow end. This means that the strongest repulsion will occur between the wide end of the stationary vane and moving vane. The repulsion at the narrow end of the stationary vane will not be as great. Therefore, the movable vane is forced to rotate towards the tapered end of the stationary vane and turn against the tension provided by the springs. This in turn causes the pointer to deflect up scale. The higher the current through the meter movement, the greater the deflection of the pointer.

Electrical Characteristics. Although moving-vane meters are primarily used to measure ac, they may also be used to measure dc if their scales are appropriately calibrated. This is possible because the moving and stationary vanes always repel each other no matter which direction the current is flowing.

In general, moving-vane meters require more current to produce full-scale deflection than the rectifier-type, moving-coil meters. Therefore, moving-vane meters are seldom used in low power circuits. They are more suitable for measuring the relatively high currents encountered in various types of ac power circuits.

Most moving-vane meters have a full scale accuracy of approximately ±5% but they cannot provide accurate readings over a wide frequency range. Most of these meters cannot provide accurate readings at frequencies that are much above 100 hertz and are used mostly in applications where 60 hertz ac power is involved.

The scale used with the moving-vane meter is usually nonlinear. In other words, the values on the meter's scale are not equally spaced. In fact, these meters usually provide a pointer deflection that increases with the square of the change in current. For example, suppose that 10 milliamperes of current causes the pointer to deflect a distance of 1 inch on the meter's scale. If this current increased to 20 milliamperes, or doubled, the pointer would not simply deflect a distance of 2 inches (twice as far). Instead, the pointer would deflect four times as far or 4 inches. Likewise, if the current increased to 3 times its initial value, the pointer would deflect 9 times as far (9 is the square of 3). If the input current becomes 4 times as great, the pointer would deflect 16 times its initial distance. Any meter that is calibrated in this manner is said to have a *square-law scale*.

A typical scale for a moving-vane meter is shown in Figure 2-8. Notice that the values near the zero end of the scale are more closely spaced. With this type of scale it is somewhat difficult to accurately measure low current values which fall near the zero end of the scale.

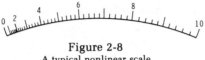

Figure 2-8
A typical nonlinear scale

Thermocouple Meters

Another type of ac meter known as a *thermocouple* meter is also commonly used to measure ac. This instrument utilizes a device known as a *thermocouple* to generate the current needed to drive the meter movement. We will now briefly examine this meter and consider its important electrical characteristics.

Meter Operation. The thermocouple meter basically consists of a *thermocouple* and a moving-coil meter movement as shown in Figure 2-9. The thermocouple consists of two dissimilar metal strips or wires which are joined together at one end. When this junction is heated, the two metals react by producing a difference of potential or voltage across their opposite ends. The thermocouple is therefore used to convert heat into an electrical voltage.

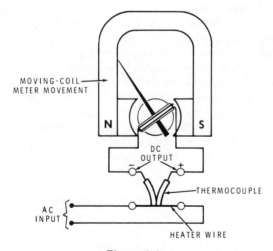

Figure 2-9
A basic thermocouple meter

71

A short *heater wire* is placed against the thermocouple's junction as shown in Figure 2-9. The ac input is applied directly to this heater wire and this wire heats up to a temperature which is determined by the amount of current flowing through the wire. The heat produced by the wire causes the thermocouple to produce a dc output voltage which in turn causes a current to flow through the moving-coil meter movement. This current causes the pointer to deflect and indicate the value of the input ac.

A higher ac input current will produce more heat within the wire, a higher dc output voltage from the thermocouple, a higher current through the meter movement, and a greater pointer deflection. Therefore, higher or lower ac input currents result in more or less deflection of the meter's pointer respectively.

Electrical Characteristics. The thermocouple meter can measure alternating currents over an extremely wide frequency range. In fact, its upper frequency limit extends well up into the radio frequency (RF) range. These meters are often used at frequencies as high as several thousand megahertz. However, they may also be used to measure dc if their scales are appropriately calibrated. This is because these meters are completely insensitive to the rate at which the input current varies. They respond only to the amount of heat that the ac or dc input can produce.

When a thermocouple meter is used to measure an alternating current that has an extremely high frequency, it is necessary to calibrate the instrument at that particular frequency. In other words, the instrument must be adjusted so that it indicates the correct ac value (usually the effective value) at that frequency. This is because a phenomenon known as *skin effect* occurs at extremely high frequencies. Skin effect occurs because high frequency ac currents tend to flow near the outer surface of a wire and this phenomenon becomes even more pronounced as frequency increases. This means that most of the conductor's interior is not used to support current and its resistance is higher than it would normally be. Therefore, the effective resistance of the heater wire and the other wires within the meter vary with frequency, thus changing the internal resistance of the meter. This change in internal resistance affects the meter's response, thus making it necessary to calibrate the meter at the frequency at which it will be used.

Thermocouple meters provide quite accurate ac measurements. Typical meters will usually have a full-scale accuracy of ±2% or ±3% and certain types of laboratory instruments may have an accuracy of ±1%. Furthermore, the thermocouple meter provides a pointer deflection which varies as the square of the change in ac input current and is just like the moving-vane meter previously described. Therefore, the thermocouple meter must also use a square-law scale.

72

Clamp-On Meters

All of the meters previously described must be connected directly to an electronic circuit or device in order to obtain a current measurement. However, there is one type of measuring instrument which does not have to be physically connected to a circuit in order to provide a current measurement. This device can be simply clamped over a conductor and it will indicate the amount of current flowing through the conductor. These clamp-on type meters are also referred to as *split-core* meters, *hook-on* meters or *snap-around* meters. We will now briefly examine one of these meters and consider its important electrical characteristics.

Meter Operation. A clamp-on meter basically consists of a transformer (which has a split core) and a rectifier-type, moving-coil meter as shown in Figure 2-10. The instrument is usually mounted within a small plastic case so that it can be held in one hand.

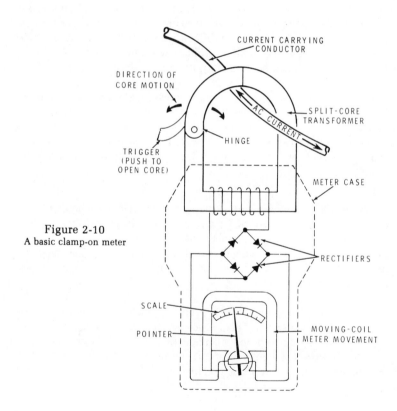

Figure 2-10
A basic clamp-on meter

73

The meter uses a split-core transformer. One side of the transformer core is hinged and is movable. This section of the core may be opened by pressing a trigger which is attached to it. To measure the ac current in a conductor, the core (which is made of soft iron) must be opened so that the conductor can be inserted inside of the core. Then the core is closed, as shown, so that it completely surrounds the conductor. It is important to insure that the core is completely closed so that no air gap is present.

The ac current flowing in the conductor produces a circular magnetic field which surrounds the conductor. The strength of this field is proportional to the current in the conductor. This magnetic field will expand and collapse as the ac increases and decreases in value and the direction of the field will change as the current changes direction. The iron core offers very little opposition to the magnetic field (much less than the surrounding air). This means that most of the lines of force will tend to flow through the core. However, in order for this to happen the magnetic lines of force must cut across the coil of wire that is wound around the opposite side of the core. When this happens, a voltage is induced into the coil which in turn causes an induced current to flow through the coil. The conductor, the core, and the coil, form a simple *transformer* with the conductor acting as the input or *primary* winding (which has only one turn) and the coil acting as the output or *secondary* winding.

The current induced in the secondary coil is an alternating current just like the current in the conductor. This ac is applied to the rectifiers which convert it to dc. The dc is then used to operate the moving-coil meter movement which causes its pointer to deflect. The meter is calibrated so that it will indicate the effective value of the ac flowing through the conductor.

Electrical Characteristics. Since the clamp-on meter depends on transformer action for operation, it can be used only to measure ac. The moving magnetic field produced by the ac in the conductor is necessary to induce a voltage in the secondary coil of the transformer. The magnetic field produced by a direct current is constant and therefore cannot pass through the transformer.

In general, the clamp-on meter is useful for measuring only relatively high alternating currents. This is because the current in the conductor must be high in order to produce a magnetic field which is strong enough to induce a significant amount of current into the secondary coil. These meters are often used to measure currents as high as several hundred amperes.

Using AC Meters

Now that you have examined some of the basic types of ac meters, it is time to consider how these meters are used in typical applications. Although there are no complicated rules governing their use, there are some important points which must be considered. The basic ac meter is essentially a current measuring device; however, this instrument may also be used to measure voltage and power. We will now see how these instruments are used to measure these various quantities and we will consider the important precautions which must be observed while using them.

Measuring Current. As mentioned previously, ac meters are essentially current measuring devices. AC meters are available for measuring alternating currents over various ranges. In general, any meter which is used to measure current (ac or dc) is referred to as an *ammeter*. However, the terms *microammeter* and *milliammeter* are also used to specifically identify meters which measure currents in the microampere and milliampere ranges.

Many ac ammeters are designed to measure current over one specific range. For example, a meter may be calibrated to measure currents from 0 to 50 milliamperes while another meter may have a scale which extends from 0 to 100 milliamperes. These two meters could be used to measure currents as high as 50 and 100 milliamperes respectively. If these values are exceeded by a substantial margin (sometimes even momentarily) it is possible that the sensitive meter movements could be damaged.

The amount of current required to deflect the meter's pointer to its full scale position is called the *meter sensitivity*. The two meters just mentioned would therefore have meter sensitivities of 50 and 100 milliamperes respectively.

Certain types of ammeters are designed to have more than one current range. Such meters are usually equipped with a range switch which effectively changes the sensitivity of the meter and its operating range. A typical multirange ammeter circuit is shown in Figure 2-11A. Notice that the instrument uses a 1 mA (1 milliampere) ac meter movement, three resistors, and a range switch. When the range switch is in the 1 mA position, only the meter movement is connected to the input terminals, thus giving the instrument a current range that extends from 0 to 1 milliampere.

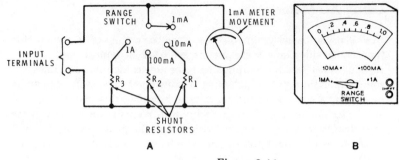

Figure 2-11
A typical multirange ammeter

When the range switch is moved to the 10 mA position, a resistor (designated as R_1) is connected in parallel with the meter movement. When the meter is in this position it can measure current values up to 10 milliamperes even though the meter movement can only handle 1 milliampere of current. This is because resistor R_1 shunts the additional current around the meter movement. In other words, when 10 milliamperes of current flow through the instrument, R_1 must carry 9 milliamperes of the current so that only 1 milliampere will flow through the meter movement.

When the range switch is moved to the 100 mA position, the instrument can measure up to 100 milliamperes of current because resistor R_2 shunts up to 99 milliamperes of current around the meter movement. When the range switch is set to the 1 A position, the instrument can measure up to 1 ampere (1000 milliamperes) of current since resistor R_3 shunts up to 999 milliamperes of current around the meter. Resistors R_1, R_2 and R_3 are appropriately referred to as *shunt resistors*.

The complete ac meter is shown in Figure 2-11B. Notice that only one scale is used and this scale is calibrated from 0 to 1. This same scale is used on each range and it is necessary to mentally adjust the decimal points on the scale so that the values correspond to the range that is used. Some meters have more than one scale, thus making it possible to read each scale directly.

When using an ammeter to measure current, you should be sure that the current that you are measuring is not beyond the range of your ammeter. Always play it safe and use the highest possible range if you are in doubt, then work down to a range which gives a measurable reading. Also, you must be sure that your meter is properly calibrated and you should always check the pointer to be sure that it is on zero when the meter is not being used. If it is not, then you must adjust the meter for a zero reading.

When using any type of ammeter, except the clamp-on type, to measure ac current, the meter must always be connected in series with the current to be measured. This means that it is necessary to break the circuit under test so that the ammeter can be inserted. For example, if you wish to measure the current flowing through a resistor, it would be necessary to break the circuit on one side of the resistor and insert the ammeter as shown in Figure 2-12A. With the ammeter in series, a complete circuit is again formed and the current must flow through the meter. When measuring the current through a component that is connected in series or parallel with other components, extreme caution must be exercised to be sure that the ammeter is in series with the correct component. For example, if you wish to measure the current through R_2 in Figure 2-12B, you must be sure that the meter is in series with R_2. If the circuit was broken and the meter was installed at points A or C, you would measure the total current which is flowing through R_1 or if it was inserted at point B you would measure the current through R_3.

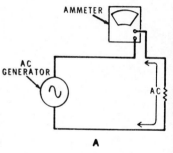

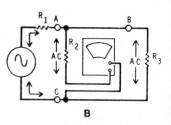

Figure 2-12
Measuring current with an ammeter

Since the alternating current periodically changes its direction, it makes no difference which direction the meter is connected. In other words, the wires connected to the meter's input terminals (shown in Figure 2-12) could be reversed and the meter would still function properly. It is only necessary to insure that the meter is in series with the alternating current you wish to measure. In this respect, ac meters are much easier to use than dc meters. The dc meters must be connected so that the current flows through them in the proper direction or in other words it is necessary to observe polarity.

Measuring Voltage. Although ac meters are essentially current measuring instruments, they may also be used to measure voltage. When used in this manner additional components are required to limit the current flowing through the meter to the proper value. Instruments that are used to measure either ac or dc voltage are referred to as *voltmeters*.

Some voltmeters are designed to measure one specific range of voltages that extend from zero to some maximum value. However, many voltmeters are capable of measuring voltages over several ranges. These multirange instruments are more versatile and are therefore more pratical in applications where a wide range of voltage values must be measured.

A typical multirange voltmeter circuit is shown in Figure 2-13A. Notice that the instrument uses a meter movement, three resistors and a range switch. When the range switch is set to the 1 V position, the instrument is capable of measuring ac voltages from 0 to 1 volt. When the range switch

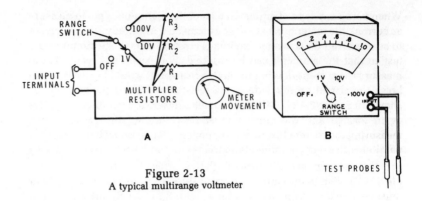

Figure 2-13
A typical multirange voltmeter

is in this position, a resistor (R_1) is connected in series with the meter movement. This resistor has a resistance value that will limit the current flowing through the meter movement to its full-scale value when the ac voltage across the input terminals has an effective value of 1 volt. The internal resistance of the meter movement itself (although it is usually very low) is also taken into consideration when this resistor value is determined. This is necessary in order to insure that the total resistance of the meter circuit will have proper value.

When the range switch is set to the 10 V position, the meter can measure effective voltage values up to 10 volts. At this time resistor R_2 is in series with the meter movement. This resistor has a higher value than R_1 and it limits the current through the meter movement to its full- scale value when the input voltage is equal to 10 volts. When the range switch is set to the 100 V position, R_3 is switched into the circuit. This resistor is larger than R_2 and it limits the current through the meter movement to its full-scale value when the input voltage is equal to 100 volts. Resistors R_1, R_2, and R_3 effectively multiply or extend the meter's voltage range by a factor of 10 in each case and these resistors are commonly referred to as *multiplier resistors.*

The complete ac voltmeter is shown in Figure 2-13B. Notice that only one scale is used for all three ranges. The instrument is also equipped with test probes which are used to connect it to a circuit. The test probes are simply placed in contact with the circuit under test and they allow voltage measurements to be made very quickly and efficiently.

When a voltmeter is used to measure a voltage, the meter must always be connected in parallel (across) the voltage source. It is not necessary to break the circuit to perform the voltage measurement. For example, if you wish to measure the voltage produced by the ac voltage source shown in Figure 2-14A, you would simply connect the meter in parallel with the source as shown. Since a resistor is also connected across the voltage source, the voltmeter is effectively measuring the voltage across the resistor as well as the source. If two resistors were connected across the voltage source and you wished to measure the voltage across just one resistor, you would connect your meter in parallel with the single resistor as shown in Figure 2-14B. In either case your voltage measurement is made by simply connecting (touching) the test probes to each side of the component to observe the voltage across the component.

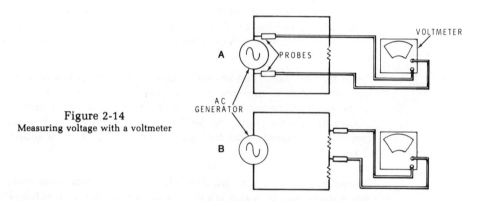

Figure 2-14
Measuring voltage with a voltmeter

When measuring an ac voltage, it makes no difference which direction the meter is connected. The test probes shown in Figure 2-14 could be reversed and the voltage readings would not change. The ac voltage continually changes direction or polarity thus making it unnecessary to observe polarity.

Always be sure that the voltage you are going to measure is within the range of the meter you are using. If in doubt, use the highest range possible and work your way down to a lower range that gives an accurate reading. Also, you should always be sure that your meter is properly adjusted for a zero reading before it is used.

Measuring Power. In ac circuits electrical power can be measured by using special types of ac meters. These instruments measure the effective values of the current and voltage involved and internally compute the amount of power and indicate its value on a calibrated scale. Electrical power is simply the rate at which electrical energy is used and it is

79

measured in units called watts. The calculation of power in ac circuits is somewhat more complex than in dc circuits. In dc circuits, the power consumed by a component (in the form of heat) is determined by multiplying the current through the component by the voltage across the component. This relationship can be shown mathematically as:

$$P = I \times E$$

Using this equation, the power (expressed in watts) is equal to the current (in amperes) times the voltage (in volts).

This same equation is also applicable to ac circuits as long as these circuits contain only resistance. In a purely resistive circuit the product of I and E will always equal P. However, ac circuits often contain other properties such as inductance and capacitance as you will learn later in this course. When these properties (which are usually present to some degree) exist, the product of I and E will not provide a true indication of power. The resulting power can appear to be much higher than it actually is.

An instrument which can be used to measure ac or dc power is called a *wattmeter*. Most wattmeters utilize a special type of meter movement called an *electrodynamometer* movement. This type of instrument will always measure the true ac power that is consumed by a component or circuit. In other words it can distinguish the true power from the apparent power.

A simple wattmeter is shown in Figure 2-15. Notice that the meter utilizes two stationary coils which are in series and a moving coil. When this instrument is used to measure the ac power consumed by a load, its

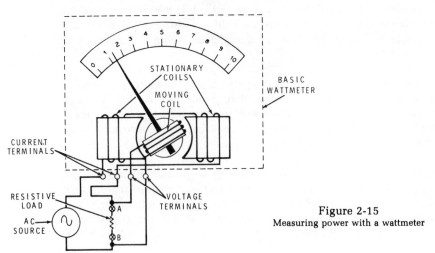

Figure 2-15
Measuring power with a wattmeter

80

stationary coils are connected in series with the load in order to measure the load current. These stationary coils are wound around two iron cores. The iron cores become magnetized when current flows through these stationary coils and they perform essentially the same functions as the permanent magnet pole pieces that were used in the moving-coil meter movement previously described. The moving coil is similar to the type used in the moving-coil meter movement. This coil is connected in parallel with the load in order to measure the voltage across the load. When the stationary and moving coils are properly connected as shown in Figure 2-15, they both produce magnetic fields which interact and cause the moving coil and its attached pointer to deflect.The amount of deflection is proportional to the product of the load voltage and the load current. Therefore, the meter actually monitors the voltage and the current and provides a scale reading which is calibrated in watts (units of power).

Most wattmeters provide reasonably accurate measurements and some are accurate to within ±1%. These meters, like the ammeters and voltmeters previously described, have internal resistances and therefore consume a certain amount of power themselves. Some wattmeters are calibrated to take this internal power loss into consideration while others are not. When using meters which do not compensate for their own loss, it is necessary to subtract the power used by the meter from the meter's reading. Manufacturers of these instruments may indicate how much power is used by the meter so that this calculation can be performed.

When using an instrument that is not compensated and the manufacturer does not indicate the internal power loss, a simple test can be performed to determine the meter's internal loss. This is done by simply disconnecting the load at points A and B as shown in Figure 2-15. When this is done, the stationary and moving coils are effectively in series and the meter measures its own internal power loss which occurs in the resistance of its coil windings. This internal power reading should be subtracted from the reading obtained with the load connected in order to obtain the actual power dissipated by the load.

The stationary and moving coils in a wattmeter have specific current and voltage ratings respectively. Therefore, when using a wattmeter you must be sure that you do not exceed the individual maximum current and voltage ratings of the device. Also, you must be careful not to exceed the maximum power that the device is designed to measure. If either of these ratings are exceeded, the instrument can be permanently damaged.

OSCILLOSCOPES

The ac meters previously described can provide reasonably accurate measurement of current and voltage, but these instruments do not allow you to see what the ac quantities actually look like. They are usually calibrated to indicate the effective or rms value of a sine wave and if they are used to measure nonsinusoidal waveforms their scales do not provide true rms readings.

When troubleshooting or analyzing electronic equipment it is often necessary to know exactly what an ac waveform looks like. In many cases, it is necessary to know its peak and peak-to-peak values, its instantaneous values, and also its frequency and period. These measurements as well as others can be performed by using a device known as an *oscilloscope*.

The oscilloscope may be used to analyze any type of ac waveform and measure its most important electrical characteristics. The oscilloscope, or *scope* as it is commonly called, is one of the most important test instruments for use in measuring ac quantities, but this device may also be used to measure dc quantities as well.

We will now examine the basic operation of the oscilloscope and then consider some of the ways it is used to measure and analyze ac waveforms. Although this discussion will be very brief, it contains background information which should prove highly beneficial.

Oscilloscope Operation

Oscilloscopes do not contain moving parts like the ac meters previously described. They are electronic test instruments which utilize various types of electronic circuits. The early oscilloscopes used vacuum tubes as the principle controlling elements in their electronic circuits. However most modern oscilloscopes now use semiconductor components such as transistors and solid state diodes.

An oscilloscope is capable of measuring an ac or dc voltage and displaying the voltage in a graphical manner. The ac or dc voltage appears as a picture on a screen which is similar to the type of screen used in a television set. The oscilloscope contains a number of controls which are used to adjust the size and the number of complete waveforms (in the case of ac) that are displayed. Most oscilloscopes are calibrated so that the waveform presented on the screen can be visually analyzed and its most important characteristics can be determined.

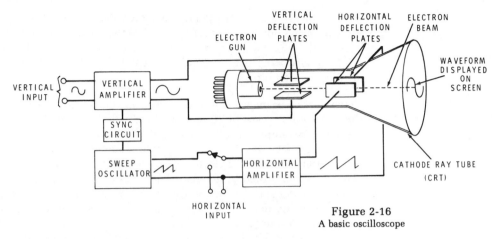

VERTICAL
DEFLECTION
PLATES

HORIZONTAL
DEFLECTION
PLATES

ELECTRON
BEAM

ELECTRON
GUN

WAVEFORM
DISPLAYED
ON
SCREEN

VERTICAL
INPUT

VERTICAL
AMPLIFIER

SYNC
CIRCUIT

SWEEP
OSCILLATOR

HORIZONTAL
AMPLIFIER

HORIZONTAL
INPUT

CATHODE RAY TUBE
(CRT)

Figure 2-16
A basic oscilloscope

A simplified block diagram of an oscilloscope is shown in Figure 2-16. The device has two input terminals which are used to measure an ac or dc voltage. These terminals must be connected in parallel with the voltage source that is to be measured and they are generally referred to as the *vertical input terminals*. The ac voltage at these terminals is applied to an amplifier circuit which increases the amplitude or magnitude of the voltage before it is applied to a device known as a *cathode ray tube* or CRT. The CRT is the device which graphically displays the ac waveform being measured.

As shown in Figure 2-16, the CRT contains an electron gun and two sets of deflection plates. These components are mounted inside of a large glass tube which fans out at one end to form a screen which closely resembles the screen on a television picture tube. The air is pumped out of the tube and the end is sealed so that the components will operate within a vacuum. In this respect the device is similar to an ordinary vacuum tube. The electron gun produces a stream of electrons which are focused into a narrow beam and aimed at the CRT screen. When the beam strikes the screen, it illuminates a phosphorus coating on the screen so that a spot of light is produced. This electron beam must also flow between the two sets of deflection plates.

The ac voltage from the vertical amplifier is applied across the *vertical deflection plates*. This alternating voltage causes the plates to become positively and negatively charged and the polarity of these charges is continually reversed. The electrons in the beam are negatively charged and tend to deflect toward the positive plate and away from the negative plate, thus causing the electron beam to bend. Since the charges on the vertical plates continually change direction, the electron beam is deflected up and down thus causing a vertical trace to appear on the CRT screen. The height of this vertical trace will depend on the amplitude of the ac voltage being measured and the amount of amplification provided by the amplifier circuit.

If the electron beam was simply moved up and down, only a vertical line or trace would appear on the screen. Such a display could indicate the peak-to-peak amplitude of a waveform but still would not indicate the exact shape of the waveform. In order to show how the waveform varies, it is necessary to move the electron beam horizontally across the screen. This is accomplished by a circuit known as a *sweep oscillator*. This circuit generates an ac sawtooth waveform which is then amplified by a *horizontal amplifier* and then applied to the *horizontal deflection plates*. The sawtooth voltage increases at a linear rate from a negative peak value to a positive peak value and then almost instantly changes back to a negative value again. The positive and negative charges on the horizontal deflection plates vary in the same manner thus causing the electron beam to move from left to right across the screen at a linear rate and then immediately jump back to the left side and start over again.

If only the sawtooth waveform was applied to the horizontal plates (no voltage on the vertical plates), then only a horizontal trace would appear on the screen. Such a trace should be thought of as a *horizontal time base* upon which the vertical signal can be made to ride. The beam simply moves from left to right in a specific period of time and then repeats this action again and again.

When the vertical ac voltage and the horizontal sawtooth voltage are both applied to the CRT, an ac waveform can be produced. As the beam moves from left to right at a linear rate with respect to time, the vertical ac voltage causes the beam to move up and down in accordance with the variations in ac voltage. If the time required for the beam to move across the screen from left to right is equal to the time required to generate one cycle of the ac input voltage, one cycle of the ac waveform will appear on the screen.

The relationship between the input ac sine wave, the sawtooth wave, and the displayed waveform is further illustrated in Figure 2-17. Notice that one complete cycle of the sine wave (Figure 2-17A) must occur in the time required to generate one complete cycle of the sawtooth wave (Figure 2-17B) in order to display one complete ac sine wave as shown in Figure 2-17C. In other words, the frequency (number of cycles per second) of the input ac waveform must be equal to the frequency of the sawtooth waveform. Furthermore, the sine wave and sawtooth wave must begin

84

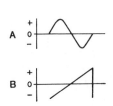

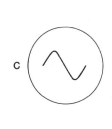

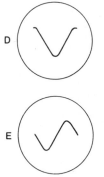

Figure 2-17
The input ac signal (A)
and the sawtooth waveform (B)
must have the same frequency for
one cycle to be displayed (C)

their cycles at the same time in order to display a sine wave that is properly oriented as shown in Figure 2-17C. If the two waveforms are not properly synchronized, the displayed waveform might appear as shown in Figure 2-17D, or Figure 2-17E. Although these are complete ac cycles, they are not properly oriented.

To insure that the input ac waveform and the sawtooth waveform are properly synchronized, a synchronization or *sync* circuit is included in the oscilloscope circuit. This circuit samples the incoming ac signal and produces a control signal which is applied to the sawtooth oscillator so that the sawtooth begins its cycle at the proper time.

The input ac waveform and sawtooth waveform do not simply occur just once. These waveforms must occur repeatedly in order to produce a picture on the screen. In other words the electron beam follows the pattern of the waveform again and again and this results in a constant picture on the screen. The phosphor on the screen produces light for only an instant after the electron beam strikes it and moves on. Therefore, this constant repetition is necessary to produce a pattern that is constantly illuminated.

Using the Oscilloscope

The oscilloscope can be used to observe various types of ac waveforms and it can be used to measure important ac values. We will now briefly consider some of the ways in which this important test instrument can be used.

Measuring Voltage. Since the oscilloscope displays an entire ac waveform, it can be used to determine instantaneous values as well as peak and peak-to-peak values.

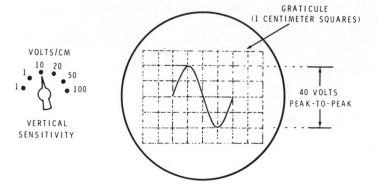

Figure 2-18
Measuring voltage with an
oscilloscope

A typical oscilloscope display is shown in Figure 2-18. Notice that the screen of the oscilloscope is marked with vertical and horizontal lines which form squares. This grid pattern is commonly referred to as a *graticule*. The squares are usually 1 centimeter high and 1 centimeter wide and are used in much the same way as a sheet of graph paper.

When observing an ac waveform as shown in Figure 2-18, the amplification of the vertical amplifier can be controlled so that the vertical height of the waveform can be adjusted. Furthermore, the vertical amplifier controls are usually calibrated so that a given input voltage will produce a specific amount of vertical deflection on the screen. The vertical amplification is usually adjusted by a control which is known as a *vertical sensitivity* control or *vertical attenuator* control. For example, suppose the vertical sensitivity control was set to the 10 volts per centimeter (cm) position as shown in Figure 2-18. This would mean that each centimeter of vertical height or deflection would represent 10 volts at the vertical input terminals. The waveform being observed in Figure 2-18 is 4 centimeters (4 squares) high and therefore has a peak-to-peak amplitude of 4 times 10 or 40 volts. The peak value of the waveform would therefore be equal to one-half of 40 or 20 volts.

In a similar manner the value at any point on the waveform can be determined by simply referring to the squares on the graticule. Also, the vertical sensitivity control can be set to other positions to either increase or decrease the sensitivity of the oscilloscope. When this is done, the squares represent other values of voltage which are either lower or higher than 10 volts.

Since the oscilloscope is capable of displaying and measuring the overall voltage of a waveform, the device is often referred to as a *peak-to-peak* measuring instrument.

Some of the less expensive oscilloscopes are not calibrated as shown in Figure 2-18. When using an oscilloscope of this type, it is necessary to apply a known dc or ac voltage to the vertical terminals and adjust the vertical amplifier for a specific amount of deflection. In other words, you calibrate the oscilloscope with a known voltage. Then you may apply the unknown voltage and determine its value.

Measuring the Period. The oscilloscope may also be used to measure the period of an ac waveform. The period (time for 1 cycle) is determined by observing the horizontal width of the waveform displayed on the screen. The oscilloscope's sawtooth oscillator can usually be adjusted so that the electron beam will move from left to right across the screen at a specific speed. The time required for the beam to move horizontally across the screen is usually referred to as the *sweep time*. The sweep time can usually be adjusted by a suitable control that is mounted on the oscilloscope. This control usually sets the amount of time (in seconds, milliseconds, or microseconds) required for the trace to move horizontally a distance of 1 centimeter.

Assume that the oscilloscope's sweep time control is set to the 5 milliseconds per centimeter position as shown in Figure 2-19. This would mean that each centimeter of horizontal deflection would represent a time interval of 5 milliseconds. The waveform being displayed in Figure 2-19 is 4 centimeters (4 squares) wide. In other words, one complete cycle occupies 4 centimeters of the trace. Therefore, the period of the waveform (time for 1 cycle) is equal to 4 times 5 or 20 milliseconds.

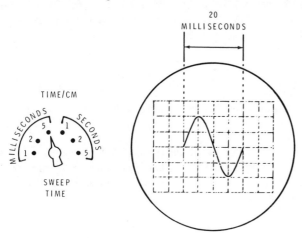

Figure 2-19
Measuring the period of an ac waveform

The sweep time control can be set to various positions so that the oscilloscope can be used to measure waveforms that have very long and very short periods. In many cases it is desirable to display only one cycle of the waveform as shown. However, the oscilloscope is capable of displaying any number of complete ac cycles on its screen. This is done by adjusting the sweep time control so that the time required to produce one complete sweep of the trace across the screen is equal to the time required to produce any given number of input ac cycles. In other words, if 4 cycles of the input ac cycle occur during the time required to generate one cycle of the sawtooth waveform (produced by the sweep generator), then 4 cycles of the input waveform will be displayed.

An oscilloscope may therefore display a number of input ac cycles. When a number of cycles are displayed on the screen, it is important to remember that it is only necessary to determine the time for one cycle in order to determine the period of the waveform.

Measuring Frequency. The frequency of an ac waveform can be determined by first measuring its period and then calculating the frequency. As explained in the previous unit, the frequency of an ac waveform (in hertz) is equal to 1 divided by the period (in seconds) and is expressed mathematically as:

$$f = \frac{1}{T}$$

For example, the waveform in Figure 2-19 has a period (T) of 20 milliseconds (0.02 seconds). This waveform would therefore have a frequency (f) of

$$f = \frac{1}{0.02} = 50 \text{ Hz}$$

The frequency would therefore be equal to 50 hertz (50 cycles per second).

There are also other ways in which the oscilloscope can be used to measure frequency without the necessity of first determining the period. However, these techniques are somewhat more complex and will not be covered at this time.

Measuring Phase Relationships. In some cases it is necessary to compare two ac waveforms of the same frequency and determine if the two waveforms coincide or occur at the same time. In many cases, two ac waveforms within the same circuit will be displaced in time or by a given number of degrees.

If two ac waveforms coincide so that their instantaneous values both occur at the same time, they are said to be *in phase* with each other. When the two waveforms are displaced, or in other words do not occur at the same time, they are said to be *out of phase*. The amount of phase displacement is usually measured in degrees. For example, the ac sine wave shown in Figure 2-20B is in phase with the waveform in Figure 2-20A. The waveform in Figure 2-20C is 90° out of phase with the waveform in Figure 2-20A and the waveform in Figure 2-20D is 180° out of phase with the waveform in Figure 2-20A.

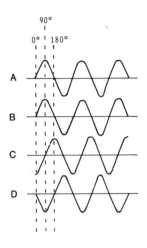

Figure 2-20
Phase relationships between ac sine waves

In order to compare the two ac sine waves and determine their phase relationship it is necessary to apply one sine wave to the oscilloscope's vertical deflection plates and the other must be applied to the scope's horizontal deflection plates. This is done by applying one waveform to the vertical input terminals and the other waveform is applied to a set of horizontal input terminals which are also located on the scope. A switch is usually provided (as shown in Figure 2-16) which disconnects the sweep oscillator and connects the horizontal input terminals to the horizontal amplifier.

When both ac sine waves are applied to the scope and the vertical and horizontal controls are properly adjusted, the electron beam will deflect in a manner that is determined by both ac waveforms. The resulting patterns displayed on the screen are referred to as *Lissajous patterns*. The phase relationship between the two waveforms can be determined by properly interpreting these unique patterns.

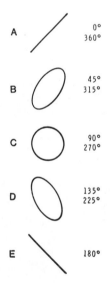

Several typical Lissajous patterns are shown in Figure 2-21. These patterns occur at phase difference intervals of 45°, starting at 0° and extending to 360°. The pattern shown in Figure 2-21A occurs when both sine waves are in phase or in other words have a phase difference of 0°. This pattern is simply a diagonal line which extends from the lower left hand portion of the screen to the upper right hand portion. The pattern in Figure 2-21B occurs when the two waveforms are 45° out of phase. It is simply an oval or an ellipse. At a phase difference of 90°, a perfect circle is formed as shown in Figure 2-21C. Then at 135° and 180° an ellipse and diagonal line are again formed as shown in Figure 2-21D and Figure 2-21E. However, these last two patterns are slanted in the opposite direction.

Figure 2-21
Typical Lissajous patterns

Any phase differences other than the values shown in Figure 2-21, will produce elliptical patterns that have shapes that are midway between the various shapes shown. It is important to realize that the Lissajous pattern changes from a diagonal line, to an ellipse, to a circle, then to an ellipse, and finally back to a diagonal line as the phase difference increases from 0° to 180°. For phase differences between 180° and 360°, the sequence is repeated but in the opposite direction as shown.

Through the use of Lissajous patterns, the oscilloscope becomes a reasonably accurate phase measuring device. However, the accuracy of the measurements taken, will depend largely on the skill of the individual using the scope. Also, the waveforms must be sinusoidal in shape in order to produce the patterns shown in Figure 2-21. Nonsinusoidal waveforms will produce irregularly shaped patterns which can be extremely difficult to analyze.

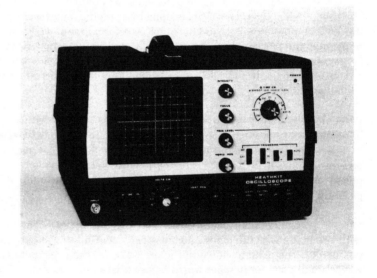

High quality oscilloscopes like the unit shown above are used extensively by service technicians and engineers. The unit shown can be used to analyze ac waveforms that have frequencies as high as 5 megahertz.

RESISTANCE IN AC CIRCUITS

Now that you have examined some of the basic test instruments which are used to measure ac values, it is time to analyze some fundamental ac circuits and the rules which apply to these circuits.

Our analysis at this time will cover only those ac circuits which contain resistance. We will consider the relationship between current, resistance, and voltage in these purely resistive circuits and compare them with similar dc circuits. Later in this course you will examine more complex ac circuits which contain additional properties such as inductance and capacitance. However, an understanding of these simple resistive circuits is necessary before you can move to these more advanced studies.

Basic AC Circuit Calculations

A simple ac circuit can be formed by simply connecting a load resistance across an ac voltage source as shown in Figure 2-22A. Although a resistor is used in this circuit, a resistive component such as a lamp or heating element would have the same effect. Such devices are, for all practical purposes, purely resistive and have a negligible amount of inductance or capacitance.

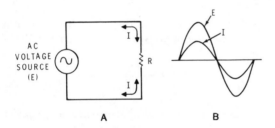

Figure 2-22
Current and voltage in a resistive ac
circuit

The ac voltage source could be an ac generator or an electronic circuit which produces an ac voltage. When this ac voltage is applied across the resistor, a corresponding ac current is produced which flows through the resistor. This current varies in amplitude and direction in accordance with the ac voltage. In other words the current is zero when the voltage is zero and it is maximum when the voltage is maximum. When the voltage changes its polarity, the current changes its direction. The voltage and current in a purely resistive ac circuit are said to be *in phase*.

The in-phase relationship between the voltage and current is shown graphically in Figure 2-22B. This figure shows that the voltage (E) waveform and the current (I) waveform pass through zero and maximum at the same time and both change direction at the same time. The two waveforms do not have exactly the same peak amplitude because they represent different quantities and are measured in different units. They are drawn together only to show that they occur simultaneously.

The value of the current flowing through the resistor in Figure 2-22A, at any given instant, depends on the voltage at that instant and the circuit resistance (which remains constant). The current at any instant can be determined by using Ohm's law in the same way that it is used in a dc circuit. In other words, the same rules and laws that apply to dc circuits, may also be used in ac circuits that contain only resistance.

Ohm's law states that the voltage, current, and resistance are mathematically related as follows:

$$E = I \times R$$

This equation states the voltage (E), which is measured in volts, is equal to the current (I), which is measured in amperes, times the resistance (R) which is measured in ohms.

This basic equation can be rearranged to show that current is equal to voltage divided by resistance or expressed mathematically:

$$I = \frac{E}{R}$$

It may also be rearranged to show that resistance is equal to voltage divided by current or expressed mathematically:

$$R = \frac{E}{I}$$

When working with ac circuits, instantaneous values of voltage and current are seldom used in ac calculations. In most cases the effective values of these ac quantities are used instead. As explained in the previous unit, the effective value of an ac voltage or current sine wave is equal to 0.707 times its peak value. An ac current sine wave with an effective value of 1 ampere effectively produces the same amount of heat (in a given resistance) as a dc current of 1 ampere. Therefore, when we use effective values we are expressing the ac quantity in terms of its dc equivalent value.

Ohm's law can be used with effective values just as easily as with instantaneous values. In other words the effective value of E will be obtained if the effective value of I is multiplied by R. Likewise the effective value of I is determined when the effective value of E is divided by R. Also, R is equal to the effective value of E divided by the effective value of I.

Consider a typical circuit which has an ac voltage source that has an effective value of 100 volts and a resistance of 100 ohms as shown in Figure 2-23. According to Ohm's law, the effective value of current (I) must be equal to:

$$I = \frac{E}{R} = \frac{100}{100} = 1 \text{ ampere}$$

Therefore an effective voltage of 100 volts will produce an effective current of 1 ampere through a resistance of 100 ohms.

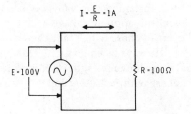

Figure 2-23
Finding the effective
value of ac current

Let's consider another situation where the resistance and current are known and the voltage must be determined. For example, suppose that the current in the circuit has an effective value of 3 amperes and the resistance is equal to 50 ohms as shown in Figure 2-24. According to Ohm's law, the applied voltage must have an effective value of

$$E = IR = (3)(50) = 150 \text{ volts}$$

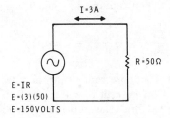

Figure 2-24
Finding the effective value of
ac voltage

Series AC Circuit Calculations

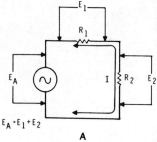

$E_A = E_1 + E_2$

A

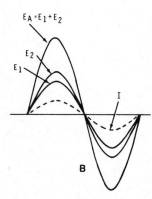

$E_A = E_1 + E_2$

E_2

E_1

I

B

Figure 2-25
Current and voltage in a series circuit

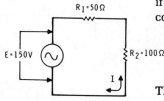

$R_1 = 50\,\Omega$

$E = 150V$ $R_2 = 100\,\Omega$

I

Figure 2-26
Calculating current and voltage
values in a series circuit

The current in a resistive circuit is always in phase with the applied voltage even when more than one resistor is used. For example, the single resistor shown in Figure 2-22A could be replaced by two series resistors so that a simple series circuit is formed as shown in Figure 2-25A. The current (I) flowing through this circuit will be limited by the total resistance of the circuit which is equal to the sum of the two resistances. This current will have the same value, at any given instant at all points in the circuit. Also, this current will be in phase with the applied voltage. This in-phase relationship between the applied voltage (E_A) and the circuit current (I) is shown in Figure 2-25B.

Since the current (I) flows through both resistors (R_1 and R_2), a voltage is dropped across each resistor. The voltage across each resistor, at any given instant, is equal to the product of the current (at that instant) and the resistance according to Ohm's law. These two voltages (designated as E_1 and E_2) are in series and their instantaneous values, when added together at any instant, will equal the applied voltage at that specific instant. In other words, E_1 and E_2 are in phase with E_A and their combined instantaneous values ($E_1 + E_2$) will always be equal to E_A. This in-phase relationship between E_A, E_1, and E_2 is shown in Figure 2-25B. Notice that all three of these voltages are in phase with each other and they are also in phase with the circuit current.

As explained earlier, it is common practice to use effective values of current and voltage when analyzing ac circuits. To illustrate this point, consider a typical series ac circuit. For example, suppose that the ac voltage source has an effective value of 150 volts and that R_1 and R_2 have values of 50 ohms and 100 ohms respectively as shown in Figure 2-26. The total resistance of the circuit (R_T) would be equal to $R_1 + R_2$. Therefore, R_T would equal 50 + 100 or 150 ohms. The circuit essentially acts as if it contained a single 150 ohm resistor as far as circuit current is concerned. Therefore, the current must be equal to E_A divided by R_T or

$$I = \frac{E_A}{R_T} = \frac{150}{150} = 1 \text{ ampere}$$

This current value of 1 ampere represents the effective value of the current in the circuit. This current flow through R_1 produces a voltage drop which according to Ohm's law, must be equal to the current times the resistance of R_1 or

$$E = IR_1 = (1)(50) = 50 \text{ volts}$$

94

This same current also flows through R_2 and produces a voltage across this resistor that is equal to:

$$E = IR_2 = (1)\ (100) = 100 \text{ volts}$$

Notice that these voltage drops are proportional to the resistance values. Also, these voltages are expressed in effective values. The sum of these two effective voltage values should be equal to the effective value of the applied voltage. This can be expressed mathematically as:

$$E_A = E_1 + E_2 = 50 + 100 = 150 \text{ volts}$$

The example just given shows how effective ac values are used in a simple series ac circuit. As you can see, these effective values are used in the same way that dc values are used. The same rules that apply to a series dc circuit containing only resistance, also apply to a series ac circuit which contains resistance.

Parallel AC Circuit Calculations

When two resistors are connected in parallel and an ac voltage is applied to them as shown in Figure 2-27A, the total current (I_T) supplied by the voltage source will be in phase with the applied voltage (E_A). This in-phase relationship between E_A and I_T is shown in Figure 2-27B. However, this total current divides and flows through the two resistors (R_1 and R_2). These two currents are designated as I_1 and I_2 in Figure 2-27A.

The individual currents (I_1 and I_2) are in phase with I_T and their instantaneous values add to produce I_T as shown in Figure 2-27B. Therefore, at any given instant I_T is equal to the sum of I_1 and I_2.

The applied voltage (E_A) appears across both resistors and this voltage is in phase with I_T, I_1 and I_2 as indicated in Figure 2-27B.

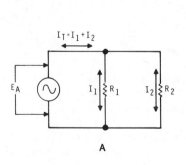

A

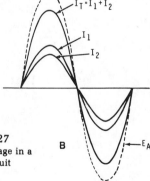

Figure 2-27
Current and voltage in a
parallel circuit

B

95

To illustrate how effective values may be used to analyze the parallel circuit in Figure 2-27, we will consider a typical parallel circuit. For example, suppose that the applied voltage is equal to 150 volts and that R_1 and R_2 have values of 50 ohms and 100 ohms respectively as shown in Figure 2-28.

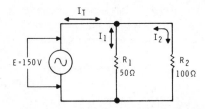

Figure 2-28
Calculating current and voltage in a parallel circuit

The current flowing through R_1 (I_1) can be determined by dividing the applied voltage (E_A) by the resistance of R_1 or

$$I_1 = \frac{E_A}{R_1} = \frac{150}{50} = 3 \text{ amperes}$$

The current through R_2 (I_2) is equal to the same applied voltage divided by the resistance of R_2 or

$$I_2 = \frac{E_A}{R_2} = \frac{150}{100} = 1.5 \text{ amperes}$$

The total current (I_T) is equal to the sum of I_1 and I_2 or

$$I_T = I_1 + I_2 = 3 + 1.5 = 4.5 \text{ amperes}$$

Since the total current is equal to 4.5 amperes and the applied voltage is equal to 150 volts, we can again use Ohm's law to determine the equivalent or total resistance of the circuit. In other words the total resistance is equal to

$$R_T = \frac{E_A}{I_T} = \frac{150}{4.5} = 33.3 \text{ ohms}$$

The circuit therefore acts as if it contained one resistor that had a value of 33.3 ohms. If such a resistor were connected across the voltage source, the same total current would result.

All of the current and voltage values just used were effective values. They were used in exactly the same way that dc current and voltage values would be used in a dc parallel circuit.

96

Power in AC Circuits

In an ac resistive circuit, power is consumed by the resistive component in the form of heat just as it is in a dc resistive circuit. The power used in either a dc or ac circuit is measured in units called *watts*.

In a dc circuit the power (P), which is measured in watts, is equal to the current (I), in amperes, times the voltage (E) in volts. This can be shown mathematically as

$$P = IE$$

This same relationship applies to ac circuits which contain only resistance. In other words, the current at a specific instant can be multiplied by the voltage of that instant to determine the instantaneous power. If all of the instantaneous values are multiplied for a complete cycle of voltage and current, we find that the power fluctuates in accordance with the voltage and current values.

A simple ac circuit is shown in Figure 2-29. The power consumed by the resistor in this circuit varies with the product of the current through the resistor and the voltage across the resistor as previously described. This relationship between the power, current, and voltage is shown in Figure 2-29B. Notice that the power curve or waveform does not extend below the zero axis (horizontal line). This is because the power is effectively dissipated in the form of heat, no matter which direction the current is flowing, and it is therefore assumed to have a positive value. Notice that the power reaches a peak value when E and I are maximum and it drops to zero when E and I are equal to zero.

Since the power fluctuates between a peak value and zero, the average power used by the circuit is midway between these two extremes. In other words, if we were to draw a line midway between the peak and zero values, this line would indicate the average power being used. It is this average power that is important in ac circuits. The average power is the power that is actually used.

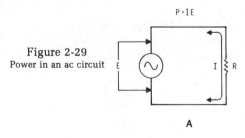

Figure 2-29
Power in an ac circuit

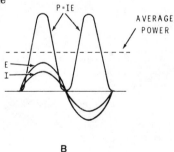

A

B

97

The average power dissipated in an ac circuit can be determined by simply multiplying the effective value of current times the effective value of voltage. Therefore, the power equation (P = IE) can be used in ac calculations as long as effective values of I and E are used. For example, suppose that the resistor in Figure 2-29 has a value of 100 ohms and the applied voltage has an effective value of 100 volts. The current in the circuit would be equal to

$$I = \frac{E_A}{R} = \frac{100}{100} = 1 \text{ ampere}$$

This effective current value of 1 ampere would be multiplied by the effective value of the applied voltage to determine the power as follows:

$$P = IE = (1)(100) = 100 \text{ watts}$$

Therefore, the resistor would consume 100 watts of power.

At times it is convenient to use another form of the power equation which expresses power in terms of voltage and resistance. This equation states that power is equal to the voltage squared divided by the resistance or stated mathematically:

$$P = \frac{E^2}{R}$$

This form of the power equation could have been used to solve the previous problem thus eliminating the additional step that was needed to first calculate the current through the resistor.

The power equation can also be expressed in terms of current and resistance as follows:

$$P = I^2 R$$

This equation simply states that power is equal to the current squared times the resistance.

These various forms of the power equation provide several means of calculating power for various combinations of known values of current, voltage, and resistance. These equations are also used in dc circuit calculations.

SUMMARY

A variety of ac meters are used to measure both ac current and ac voltage. One of the most popular types is the rectifier-type, moving-coil meter, although moving-vane meters, thermocouple meters, and clamp-on meters are also widely used.

A meter that is used to measure current is called an ammeter. Normally an ammeter must be connected in series with the current to be measured. This means that it is necessary to break the circuit under test and install the ammeter within the circuit. Only the clamp-on meter can be used without breaking the circuit under test. This meter simply clamps over the conductor that is carrying the current to be measured.

The oscilloscope is perhaps the most versatile of all test instruments. It can be used to measure the peak and peak-to-peak values of an ac waveform and also any specific instantaneous value on the waveform. This instrument may also be used to measure the period and frequency of a waveform and it can even compare ac sine waves and determine their phase relationships.

AC circuits which contain only resistance can be analyzed in much the same way as dc circuits which contain only resistance. When analyzing ac circuits, it is common practice to use the effective values of the currents and voltages involved.

Unit 3

CAPACITIVE CIRCUITS

INTRODUCTION

In this unit you are going to study capacitors and how they are used in ac circuits. Capacitors are one of the key electronic components used in alternating current circuits. When combined with resistors and/or inductors, capacitors can be used to form a wide variety of useful electronic networks.

The prerequisite for this program in AC Electronics is previous training in dc electronics. This training should have included a study of, dc principles as well as information on capacitors and inductors. This means that you should know what a capacitor is, how it works and how it is used in dc circuits. However, to refresh your knowledge, the first part of this unit is devoted to a review of capacitors and capacitance.

Next you will study the effect of capacitance in an ac circuit. The relationship between the current and voltage in a capacitive circuit will be considered in detail. Some practical applications of basic capacitive circuits will then be discussed.

The capacitor is one electronic component that you will encounter in almost all electronic circuits. For that reason, this unit is very important to your understanding of electronic circuit operation.

REVIEW OF CAPACITORS AND CAPACITANCE

Before you learn how capacitors are used in alternating current circuits, it is desirable to review the basic operation and characteristics of a capacitor. You can understand the operation of a capacitor in an ac circuit better if you are familiar with the operation of a capacitor in dc circuits. This section will help you review the key facts about the capacitors.

A capacitor is an electronic component that can store electrical energy in the form of an electric field. In its simplest form, a capacitor consists of two conducting plates separated by an insulator called the dielectric. Figure 3-1 shows a simple capacitor made up of two square metal plates separated by an air dielectric.

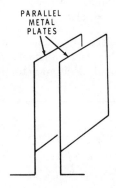

PARALLEL
METAL
PLATES

Figure 3-1
A simple capacitor

If we apply a dc voltage to the plates of a capacitor, the capacitor will become charged. Figure 3-2A shows a battery connected to the simple two-plate capacitor. The positive terminal of the battery attracts the electrons in the left-hand plate of the capacitor thereby leaving that plate with a positive charge. Electrons from the negative terminal of the battery move on to the right-hand plate giving it a negative charge. In this state, the capacitor is said to be charged. Since the positive and negative electrical charges on the plates attract one another, there will be a force field set up between the two plates. However, there is no electrical current flow through the capacitor because of the insulating dielectric between the two plates. The only time current flows is when the battery is initially connected. Current flow, or the movement of electrons, takes place only during that brief instant of time that it takes for the capacitor to charge.

Most capacitors can be charged in either direction. Figure 3-2B shows how the two-plate capacitor can be charged in the opposite direction simply by reversing the battery leads. In this state, the left-hand plate takes on a negative charge and the right-hand plate takes on a positive charge.

If the battery in the circuits of Figure 3-2 is removed from the capacitor, the electrical charges on the plates remain. The attraction of the positive and negative charges on the two plates across the dielectric holds the charges in place and the capacitor remains charged. As long as the two plates are insulated from one another, the capacitor will remain charged.

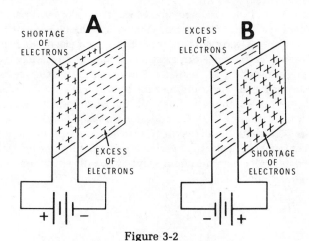

Figure 3-2
Charging a capacitor with a battery

The capacitor can be discharged by shorting the plates together by connecting a wire between them as shown in Figure 3-3. When a capacitor is shorted in this way, the electrons on the negative plate flow through the shorting connection to the positively charged plate. Momentarily, electrons will flow creating a current. The excess of electrons on the right-hand plate will neutralize the positive charge on the left-hand plate. This discharge action gives the capacitor a neutral or zero charge.

As you can see, a capacitor can store electrical energy in the form of a charge. The capacitor is charged by an external voltage source, then retains that charge because of the attraction of the opposite charges on the two plates across the dielectric. The capacitor can be discharged by connecting the two plates together. The charge is then neutralized. No current flows through the capacitor itself, but current does flow in a capacitive circuit during the time the capacitor is being charged or discharged. Figure 3-4 shows the electronic symbol used to represent a capacitor in schematic diagrams of electronic circuits.

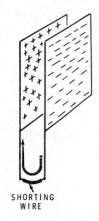

SHORTING
WIRE

Figure 3-3
Discharging a capacitor
by shorting the plates

Figure 3-4
Schematic symbol for a capacitor

The ability of an electronic component to store an electrical charge is referred to as capacitance. Components that exhibit this characteristic are known as capacitors. Most capacitors are two-leaded electronic components. However, capacitance can exist in any situation where two conductors are separated by an insulator. For example, capacitance exists between an insulated wire running near a metal chassis. The chassis represents one plate of the capacitor while the wire represents the other plate. The insulation of the wire and any air space between the two conductors represents the dielectric. Capacitance not in the form of a physical component is generally referred to as distributed or stray capacitance. It exists in all electronic circuits and can greatly effect their operation at high frequencies.

Units of Capacitance

The measure of a capacitor's ability to store an electrical charge is referred to as its capacitance. The unit of capacitance is the farad. One farad of capacitance indicates that a capacitor will store one coulomb of charge when the voltage applied to the capacitor is one volt. This relationship is expressed by the simple expression below:

$$C = \frac{Q}{E}$$

C equals the capacitance in farads, Q equals the quantity of the electrical charge expressed in coulombs (one coulomb = 6.25×10^{18} electrons), and E represents the applied voltage in volts. This expression tells us that the larger the quantity of electrons that a capacitor can store for a given applied voltage, the greater its capacitance. A large capacitor is one that is capable of storing a large charge even when a small voltage is applied. A small capacitor is one that cannot hold a great charge even though a large voltage may be applied to it. The capacitance is a function of the physical characteristics of the capacitor.

The farad is a very large unit of electrical capacitance. A one-farad capacitor would be physically very large. A capacitor of this size is much larger than is generally required in most electronic applications. Most capacitors used in electronic circuits have a capacitance significantly smaller than a farad. Typical electronic capacitors have a capacitance of one millionth of a farad or less. One millionth of a farad is called a microfarad (abbreviated μf). Another commonly used unit of capacitance is the picofarad (abbreviated pf) which is one millionth of a microfarad. Table I shows the relationship between the farad, microfarad and picofarad. Table II shows how to convert from one unit to the other.

TABLE I

Units of Capacitance

1 farad =	1,000,000 or 10^6 microfarads
1 farad =	1,000,000,000,000 or 10^{12} picofarads
1 microfarad =	.000001 or 10^{-6} farad
1 microfarad =	1,000,000 or 10^6 picofarads
1 picofarad =	.000000000001 or 10^{-12} farad
1 picofarad =	.000001 or 10^{-6} microfarad
farad =	f
microfarad =	μf
picofarad =	pf

The examples below illustrate the use of Tables I and II in converting from one unit of capacitance to another.

1. Convert 25μf to f:
   25μf = 25 $\times$ 10^{-6} = 25 $\times$ .000001 = .000025f
2. Convert 470 pf to f:
   470pf = 470 $\times$ 10^{-12} = 470 $\times$ .000000000001 = .00000000047f
3. Convert 1000pf to μf:
   1000pf = 1000 $\times$ 10^{-6} = 1000 $\times$ .000001 = .001μf
4. Convert .00082μf to pf:
   .00082μf = .00082 $\times$ 10^6 = .00082 $\times$ 1,000,000 = 820pf

TABLE II

Converting Units of Capacitance

To Convert	To	Action
Farads	Microfarads	Multiply by 1,000,000 (10^6) or move the decimal point 6 places to the right.
Farads	Picofarads	Multiply by 1,000,000,000,000 (10^{12}) or move the decimal point 12 places to the right.
Microfarads	Farads	Divide by 1,000,000 (10^6) or multiply by .000001 (10^{-6}). Move the decimal point 6 places to the left.
Microfarads	Picofarads	Multiply by 1,000,000 (10^6) or move the decimal point 6 places to the right.
Picofarads	Farads	Divide by 1,000,000,000,000 (10^{12}) or multiply by .000000000001 (10^{-12}). Move the decimal point 12 places to the left.
Picofarads	Microfarads	Divide by 1,000,000 (10^6) or multiply by .000001 (10^{-6}). Move the decimal point 6 places to the left.

Factors Affecting Capacitance

The ability of the capacitor to store an electrical charge is referred to as capacitance. The capacitance of a capacitor is determined strictly by its physical characteristics. Specifically, the capacitance is determined by the total area of the conducting plates, the distance separating the plates, and the type of dielectric used.

The larger the plates in a capacitor, the greater the charge that the capacitor can store. Larger plates can store more electrons and have more electrons to give up than smaller plates. The greater the charge (Q), the greater the capacitance (C) for a given applied voltage. Therefore, increasing the plate area of a capacitor increases its capacitance.

The spacing between the plates of a capacitor also determines the amount of charge that it can store. Decreasing the distance between the two conducting plates of a capacitor increases the intensity of the attraction of the charges on the two plates. The greater this attraction, the greater the quantity of electrons that can be stored in the capacitor for a given applied voltage. Moving the plates closer together therefore increases the capacitance. Separating the plates by a greater distance causes the capacitance to decrease. In most capacitors, the distance between the two plates is a function of the thickness of the dielectric material used in the capacitor.

The type of insulating material used between the two plates of the capacitor also has an important effect on the amount of charge that a capacitor can store. Generally, the better the insulator that the dielectric is, the greater the capacitance. Many different types of insulating materials are used as dielectrics in capacitors. Insulators such as air, oil, paper, glass and various types of plastics are widely used. Each type of insulator has a specific dielectric constant (K) that effects the amount of charge that a capacitor can store for a given applied voltage. Table III shows the dielectric constants for typical insulating materials.

TABLE III

Dielectric Constants

Material	Dielectric Constant (K)
Air	1
Paper	3.5
Mica	6
Glass	6-10

The formula below shows the relationship between the factors effecting the capacitance of a capacitor. C is the capacitance in picofarads, A is the plate area in square inches, D is the distance between the plates in inches and K is the dielectric constant. If all of these factors are known, the capacitance can be computed:

$$C = \frac{0.2248AK}{D}$$

The capacitance is directly proportional to the plate area and the dielectric constant and inversely proportional to the plate spacing.

Example: What is the capacitance of a paper capacitor with a plate area of 18 square inches and spacing of .005 inches?

$$C = \frac{.2248\,(18)\,(3.5)}{.005} = 2832.5 \text{ pf}$$

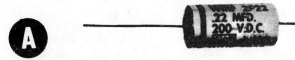

Figure 3-5
Typical fixed capacitors

(A) paper, (B) ceramic, (C) mica

109

Types of Capacitors

There are two basic types of capacitors used in electronic applications, fixed capacitors and variable capacitors. A fixed capacitor is one that has a constant capacitance. Its physical construction is such that the capacitance remains essentially at one value under all conditions. A variable capacitor, on the other hand, is one that is designed so that its capacitance can be changed.

Another method of classifying capacitors is by the type of dielectric used. The most commonly used dielectrics in capacitors for electronic applications are mica, ceramic, paper, and plastic films such as Mylar*. Oil is also used as a dielectric in some large capacitors for high power applications. Another type of capacitor widely used in electronic applications is the electrolytic. This type of capacitor uses aluminum plates where the dielectric is a thin layer of aluminum oxide that is developed during the manufacturing of the capacitor by the application of a dc voltage. The dielectric is formed by an electro-chemical process right on the plates so that plate spacing is extremely small. This permits very high values of capacitance to be contained within a small package.

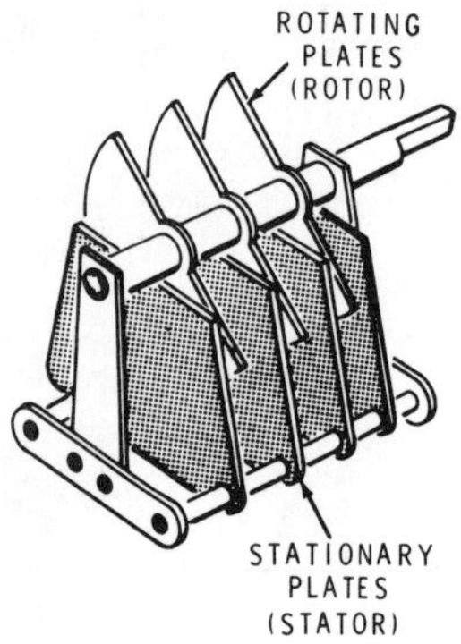

The various fixed capacitor types have been developed to meet the very broad range of needs in electronic applications. For example, mica and ceramic capacitors are excellent for very high frequency applications. Paper and plastic film capacitors are used in lower frequency applications where higher values of capacitance are generally required. Electrolytics are used where very high values of capacitance are required. Figure 3-5 shows several types of fixed capacitors used in electronic circuits.

Most variable capacitors have an air dielectric and are therefore referred to as air capacitors. Figure 3-6 shows a typical variable capacitor. Note the fixed plates (stator) and the variable plates (rotor). The capacitance increases as the variable plates are rotated into mesh with the fixed plates. Maximum capacitance occurs with full mesh, minimum capacitance occurs with no mesh.

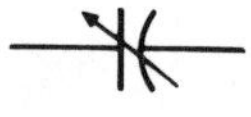

Another type of variable capacitor called a trimmer uses a mica dielectric to obtain a relatively high value of capacitance in a small size. Trimmer capacitors with a ceramic dielectric are also available. Figure 3-7 shows the symbol used to represent a variable capacitor in a schematic diagram.

*DuPont Registered Trademark

ROTATING
PLATES
(ROTOR)

STATIONARY
PLATES
(STATOR)

Figure 3-6
The variable capacitor

Figure 3-7
The schematic symbol
for a variable capacitor

Capacitor Ratings

All capacitors are rated according to two basic characteristics: capacitance and voltage. Capacitors come in a variety of standard capacitance values. Standard ranges cover from as high as several thousand microfarads to as low as one picofarad.

Another capacitor characteristic is its voltage rating. All capacitors are designed to be able to withstand an applied voltage of a certain maximum value. This voltage rating is a function of the type of dielectric and its thickness. If the voltage rating of a capacitor is exceeded, it is possible to rupture the dielectric and cause the plates of the capacitor to short together. In some very high value capacitors, the dielectric material is extremely thin. A very thin dielectric is responsible for the high capacitance value, but it is also easier to puncture the dielectric with a high voltage.

Capacitors are generally rated in terms of an operating or working voltage. This is the maximum voltage that can be applied to the capacitor on a continuous basis. For example, a capacitor with a 200-volt rating can operate continuously with any value of dc voltage less than 200 volts. Applying a voltage greater than 200 volts may cause damage to the capacitor. On the other hand the 200-volt capacitor, would operate quite satisfactorily at 50 volts dc. You will often see working voltage abbreviated WV.

Another capacitor voltage rating is peak voltage. The peak voltage is the maximum voltage that a capacitor can withstand when used with an ac voltage. For example assume that a capacitor has a peak voltage rating 1000 volts. This means the peak ac value applied to the capacitor must never exceed 1000 volts. If a 1000-volt rms sine wave were applied to the capacitor, its voltage rating would be exceeded and the capacitor may be damaged. The peak value of a 1000-volt rms sine wave is 1000×1.41 or 1410 volts. This of course exceeds the peak voltage rating.

Both the capacitance and voltage ratings of a capacitor must be considered in terms of the tolerances of those ratings. It is extremely difficult and expensive to design components which have capacitance and voltage ratings exactly as specified. Therefore, most practical electronic components have a tolerance range on their ratings. A typical capacitance value might have a tolerance of plus and minus 10% ($\pm 10\%$). This means that the actual value of the capacitance could be 10% higher or 10% lower than the actual rated value. While 10% is a typical capacitance tolerance, capacitors with wider tolerance ranges and narrower ranges are also available. One percent tolerance capacitors are available. Some very large capacitors have tolerances as great as 80%.

111

Tolerances on voltage ratings are generally undefined and very broad. The voltage ratings of a capacitor are also very conservative so that the actual voltage rating is generally much higher than the stated working value. While this is true, it is still poor practice to use a capacitor in a circuit where the voltage across it will exceed its rated value, despite the fact that its actual capability might be higher.

Capacitor Defects

Like any electronic component, capacitors are subject to defects. Defects may occur in the manufacturing process or may be caused by an improper electrical condition in the circuits in which they are used. There are four common ways that capacitors fail. A capacitor can become shorted or open. The capacitor can have excess leakage or can change in value.

A shorted capacitor is one where the plates touch or where a very low resistance path is formed between the plates. Shorts occur when the dielectric is punctured or otherwise fails. A shorted capacitor acts as a very low resistance and can often cause damage to other components in a circuit.

An open capacitor occurs when one or possibly both of the leads become disconnected from the plates. An open capacitor acts like a very high resistance or open circuit.

A capacitor with excessive leakage is one where a resistive path is formed between the two plates. A good capacitor should be a complete open circuit to direct current and should have infinite resistance between the two leads. However a resistance is sometimes developed in a capacitor when the dielectric fails. The resistance that develops between the two plates is referred to as leakage resistance. It has the effect of an external resistor connected across the capacitor leads. Leakage can generally be detected by measuring the capacitor with an ohmmeter.

Capacitors can also change in value. A fixed capacitor has a specific given value, but when used in a circuit the capacitor may not have that capacitance. Because of a defect in manufacture or improper use, the capacitor value may change. The capacitance value can change because of an incorrect applied voltage or because of an excessive temperature condition.

112

Capacitors in Series and Parallel

Capacitors are often connected in series or parallel to form new values of capacitance. When two or more capacitors are connected in series or parallel, the capacitance of the combination is greater than or less than the values of the individual capacitors. Because such combinations occur so frequently in electronic circuits, it is desirable to know how to compute the total capacitance of the various combinations. The simple procedures given below show you how to do this.

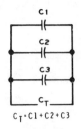

Capacitors in Parallel. When two or more capacitors are connected in parallel as shown in Figure 3-8, the total capacitance of the combination is simply equal to the sum of the individual capacitances. If the total capacitance of the combination is designated C_T the total capacitance is the sum of the individual capacitance indicated by the expression:

$$C_T = C_1 + C_2 + C_3$$

Figure 3-8
Capacitors connected in parallel

For example, if $C_1 = .015\mu f$, $C_2 = .002\mu f$, and $C_3 = 1000pf$, the total sum of the combination will be:

$$C_T = .015 + .002 + .001 = .018\mu f$$

The total capacitance is always greater than any single value in the combination.

The most important thing to note in making capacitance calculations such as this, is that all values of capacitance must be expressed in the same units. In this example, C_1 and C_2 are given in μf. C_3 is expressed in pf. In order to make the proper calculation, all values were expressed in μf. A 1,000pf capacitor is the same as a .001μf capacitor. Remember that to convert pf to μf the capacitance value in pf is divided by 1,000,000 or 10^6.

Capacitors in Series. When two or more capacitors are connected in series as shown in Figure 3-9, the total capacitance of the combination will be less than the capacitance of the smaller capacitor in the combination.

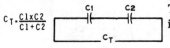

$$C_T = \frac{C1 \times C2}{C1 + C2}$$

Figure 3-9
Capacitors in series

The formula finding the total capacitance C_T of two capacitors connected in series is given below:

$$C_T = \frac{C_1 C_2}{C_1 + C_2}$$

For example, assume that two capacitors. $C_1 = .001\mu f$ and $C_2 = 390pf$, are connected in series. The total combination will be as given below:

$$C_T = \frac{1000 \times 390}{1000 + 390} = \frac{390000}{1390} = 280.6pf$$

Note that C_1 is expressed in μf while C_2 is expressed in pf. In order to compute the proper value of the capacitance, the values of both capacitors must be expressed in the same units. Here both were expressed in picofarads. A .001μf capacitor is equal to 1000pf. Note that the total capacitance of the combination is less than the smaller value of capacitance.

When more than two capacitors are put in series, the formula given above can be used to compute the total by combining the capacitors two at a time. For example, in a circuit with three capacitors in series, the combination capacitance of two of them would first be computed. This combination capacitance would then be combined with the third value of capacitor in the same formula to arrive at the correct total.

114

Capacitors in DC Circuits

When capacitors are used in circuits involving dc voltages, they exhibit certain properties and operational characteristics. When used in ac circuits, the capacitor operates in a different manner. The major part of this unit is devoted to a discussion of the operation of a capacitor in circuits involving ac voltages. However, many ac circuits also involve the use of dc voltages. For that reason, it is desirable to review the operation of a capacitor in a dc circuit.

Because there is no direct electrical connection between the plates of a capacitor, a capacitor represents essentially an open circuit or infinite resistance to a dc voltage. There is no direct conducting path for electrons to flow through the capacitor. However, when a capacitor is charged or discharged, a movement of electrons does take place within the circuit. This movement of electrons does constitute a current flow, but this charging or discharging action is transient in nature. This means that it occurs for only a brief period of time. Once the capacitor is charged or discharged, no current flows. While no electrons cross between the plates of a capacitor, the charging and discharging action does cause a movement of electrons in the external circuit. A component connected in series with the capacitor, such as a resistor, will have a voltage drop developed across it as the charging or discharging current occurs.

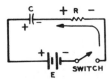

Figure 3-10

A series RC circuit

Figure 3-10 shows a simple series resistor-capacitor (RC) circuit connected to a dc voltage source. With the switch open and the capacitor initially discharged, no current flows in the circuit. However, when the switch is closed, the capacitor will charge. Electrons from the negative terminal of the battery will flow through the switch and the resistor to the right-hand plate of the capacitor. Electrons are drawn from the left-hand plate of the capacitor by the positive terminal of the battery leaving it positively charged. Once the capacitor is charged, no further current flow takes place. During the charging time, electrons will flow through resistor (R) in the direction shown. The results will be a voltage drop across this resistance with the polarity as indicated. This voltage will appear only momentarily as the capacitor is charged. When the switch is initially closed, electrons will begin to flow immediately and a peak voltage will appear across the resistor. As the capacitor charges, the voltage drop across the resistance will decrease. The voltage across the resistor will be zero when the capacitor becomes fully charged.

115

The series combination of resistance and capacitance shown in Figure 3-10 is a common configuration. There are many variations and applications of this simple circuit. The most important characteristic of this combination is the time that it takes to charge or discharge the capacitor. This time is referred to as the time constant (T). The time constant is the period of time it takes for a capacitor (C) to charge to 63.2% of the value of the applied voltage through the resistor (R). This time is the product of the resistance and capacitance value. Or the time constant is:

$$T = RC$$

The time constant (T) in seconds is equal to the capacitance in farads multiplied by the resistance in ohms. Or the time constant in seconds is equal to the resistance in megohms multiplied by the capacitance in microfarads. What is the time constant of a 25μf capacitor and a .47 megohm resistor? The applied voltage is 12 volts.

$$T = 25 \times .47 = 11.75 \text{ seconds}$$

In other words, it takes the 25μf capacitor 11.75 seconds to charge to 63.2% of 12 (.632 $\times$ 12 = 7.58) or 7.58 volts through a .47 megohm resistor.

After one time constant, the voltage across a capacitor will be equal to 63.2% of the applied voltage. It takes approximately five time constants (5T) for the capacitor to fully charge to the applied voltage. In the example above, it would take 5 $\times$ 11.75 = 58.75 seconds for the capacitor to charge to 12 volts.

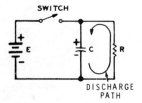

Figure 3-11
A parallel RC circuit

The time constant relationship above also holds true for the discharge of a capacitor. See Figure 3-11. If a capacitor is initially charged to a given voltage by closing the switch and then allowed to discharge through the resistor by opening the switch, the time constant is the period of time that it takes the capacitor to discharge to 36.8% of the initial charge voltage. This is another way of saying that in one time constant, the capacitor will lose 63.2% of its initial charge. It takes approximately five time constants for the capacitor to discharge completely. If the capacitor is initially charged to 36 volts, after one time constant the voltage is 36 $\times$ .368 = 13.25 volts.

CAPACITORS IN AC CIRCUITS

When an ac voltage is applied to a capacitor, alternating current flows in the circuit. Figure 3-12 shows a sine-wave generator (E) connected to a capacitor (C). As the ac voltage varies, the current in the circuit follows a sinusoidal variation. While electrons do not pass from one plate to the other through the dielectric, electrons do flow in the circuit external to the capacitor as if they did. As the applied ac voltage rises and falls, the capacitor charges and discharges.

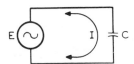

Figure 3-12

An ac voltage source connected to a capacitor.

Figure 3-13 gives a better idea about the action of electron flow in a capacitor. This illustration shows the effect of an applied ac voltage on the atoms in the dielectric of a capacitor. In Figure 3-13A, the external applied voltage is zero. At this time the electrons in the atoms of the dielectric rotate normally about their nuclei. Now, note what happens to the orbit of the electrons when the capacitor is charged as shown in Figure 3-13B. Here the upper plate is made negative with respect to the lower plate. The electrons orbiting the nucleus of the atoms in the dielectric are repelled by the negative plate and attracted by the positive plate. This distorts the orbit of the electrons. The amount of voltage applied to the capacitor determines the amount of orbit distortion. Figure 3-13C shows the orbit distortion when the capacitor is charged in the opposite direction.

When a constantly changing ac voltage is applied to the capacitor, the polarity of the applied voltage alternates causing the electrons in the dielectric to shift from one direction to the other. While the amount of shift in the electrons is small, it nevertheless constitutes a movement of electrons within the dielectric. While none of the electrons actually break loose from their orbits and flow in the external circuit, we can say that the movement of electrons constitutes a current flow. Naturally as the capacitor is charged and discharged by the ac voltage, the movement of electrons onto one plate and off the other in the external circuit represents current flow. If the applied voltage is a sine wave, the current flow in the circuit will also be sinusoidal.

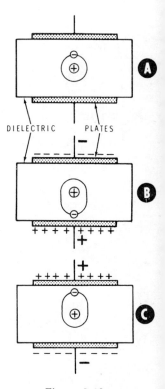

Figure 3-13

Electron distortion in the dielectric of a charged capacitor.

Current-Voltage Relationships
in Capacitive AC Circuits

The relationship between the current and the applied voltage in a capacitive circuit is different from purely resistive ac circuits. Because of the way that a capacitor works, the current and voltage in a capacitive ac circuit are not in step with one another. In a circuit where ac voltage is applied to a resistance, current through the resistor follows the voltage applied to it. We say that the current and voltage in such a circuit are in phase. The positive and negative half cycles of voltage and current in a resistive ac circuit are in step with one another.

In a capacitive ac circuit, the capacitor constantly charges and discharges with a change in the applied voltage. Once the capacitor is initially charged, the voltage across it acts as a voltage source. Its effect is to oppose changes in the external supply voltage. Since the capacitor must charge or discharge to follow the changes in the applied voltage, the resulting current flow is out of step with the changes in the applied voltage. We say that there is a phase shift between the voltage and current in the circuit.

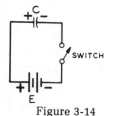

Figure 3-14
A capacitor being charged by a dc source.

To understand the relationship between the current and voltage in a capacitive ac circuit, we will go back to the basic principles of capacitor operation with a dc voltage. Figure 3-14 shows the dc voltage source (E) connected to a capacitor (C). The capacitor is initially discharged and the switch is open. When the switch is closed, the instantaneous voltage across the capacitor is zero. While it would appear that the voltage across the capacitor is initially equal to the applied voltage, the instant the switch is closed, it is zero, since electrons have not had time to flow to and from the plates of the capacitor. In other words, on the initial closure of the switch, the capacitor is still uncharged. Immediately thereafter, electrons begin to flow. Electrons flow from the negative terminal of the battery to the right-hand plate of the capacitor giving that plate an excess of electrons and a negative charge. At the same time, electrons are drawn from the left-hand plate of the capacitor to the positive terminal of the battery giving the left-hand plate a positive charge. As electrons begin to flow, a voltage builds up across the capacitor. The polarity of voltage across the capacitor is as indicated in Figure 3-14. Note that this voltage is in direct opposition to the applied voltage. When the capacitor is fully charged to the applied voltage, the two voltages equal one another and their effects cancel. The effective voltage in the circuits is zero, so no further current flows.

Figure 3-15 illustrates the relationship between the current and capacitor voltage in the circuit of Figure 3-14. When the switch is initially closed, the capacitor voltage is zero while the current in the circuit is maximum. As the electron flow charges the capacitor, the capacitor voltage begins to build up. The capacitor voltage opposes the applied voltage thereby reducing the amount of current flowing in the circuit. When the capacitor becomes fully charged, the current in the circuit is reduced to zero.

Now, using the relationship shown in Figure 3-15, we can show the actual current and voltage relationships in a capacitor circuit when an ac signal is applied. Once the capacitor becomes charged, the voltage across it follows the applied voltage exactly. However the current flowing in the circuit is out of step with the voltage as indicated in Figure 3-15.

The exact relationship between the current and voltage is a capacitive circuit when a sine-wave ac signal is applied as shown in Figure 3-16. Note that when the current is at maximum, the voltage across the capacitor is zero. As you can see, there is a phase shift between the current and voltage in the circuit. This phase shift is expressed in terms of degrees. Remember that one complete cycle of an ac sine wave contains 360 degrees. The amount of phase shift in the capacitive circuit is one-fourth of this or 90 degrees. We say that the current and voltage in a purely capacitive circuit are 90 degrees out of phase with one another. Another important fact to note is that the change in current *leads* the change in voltage. Looking at Figure 3-16 you can see that the capacitor voltage change *follows* the current change in time. We say that the current leads the voltage in a capacitive circuit. In other words, there is a 90-degree leading phase shift in a purely capacitive circuit.

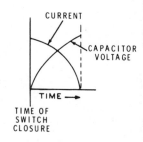

Figure 3-15

Relationship between voltage and current in a charging capacitor.

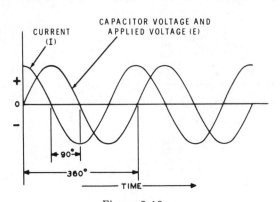

Figure 3-16

Current and voltage relationship in a purely capacitive ac circuit.

119

Capacitive Reactance

It is the basic nature of a capacitor to oppose changes in voltage. In a dc circuit, a capacitor will charge to the applied voltage. If the applied voltage is increased, the capacitor will then charge to the new higher voltage. If the applied voltage is decreased, the capacitor will discharge until the voltage across it equals the new lower applied voltage. Keep in mind that it takes a finite period of time for a capacitor to charge or discharge to a new voltage level. The charge and discharge time is a function of the capacitor size and the value of any series resistance in the circuit.

In an ac circuit, the capacitor is constantly charging and discharging. The voltage across the capacitor is in constant opposition to the applied voltage. This constant opposition to changes in the applied voltage creates an opposition to current flow in the circuit. This opposition to the flow of ac offered by a capacitor is called capacitive reactance. Capacitive reactance is represented by the symbol X_c and, like resistance, is measured in ohms.

The capacitive reactance of a capacitor is determined by its capacitance and the frequency of the applied voltage. The amount of opposition offered to current flow in an ac circuit by a capacitor is a function of the capacitance and the frequency of the ac voltage. The capacitive reactance is inversely proportional to the capacitance and the frequency. This means that increasing the capacitance or the frequency will cause the reactance to decrease. Decreasing the capacitance or frequency causes the reactance to increase.

The capacitive reactance is calculated by using the expression:

$$X_c = \frac{1}{2\pi fC}$$

In this expression, X_c is the capacitive reactance in ohms, f is the frequency in Hz, C is the capacitance in farads and pi (π) is the constant 3.14. Since $1/2\pi = 1/6.28 = .1592$, you can simplify the expression as shown below:

$$X_c = \frac{.1592}{fC}$$

120

Since farads is an unrealistically high unit of capacitance, this formula becomes somewhat difficult to manipulate. If we assume that the capacitance (C) of the formula is expressed in microfarads, the expression for capacitive reactance now becomes:

$$X_c = \frac{159200}{fC}$$

The following examples will show how to compute the capacitive reactance when the frequency of operation and the capacitance are known. These examples will also illustrate the effect of frequency and capacitance on the reactance.

1. What is the reactance of a $1\mu f$ capacitor at 60 Hz?

$$X_c = \frac{159200}{60(1)} = \frac{159200}{60} = 2653 \text{ ohms}$$

2. If the frequency is increased to 120 Hz, what is the new value of reactance of the $1\mu f$ capacitor?

$$X_c = \frac{159200}{120(1)} = 1326 \text{ ohms}$$

Increasing the frequency decreases the reactance.

3. What is the reactance of a 1000pf capacitor at 2 kHz?

$$1000pf = .001\mu f$$

$$2 \text{ kHz} = 2000 \text{ Hz}$$

$$X_c = \frac{159200}{2000 \, (.001)} = 79600 \text{ ohms}$$

4. If the capacitance is decreased to 500pf, what is the new reactance at 2 kHz?

$$500pf = .0005\mu f$$

$$X_c = \frac{159200}{2000(.0005)} = 159,200 \text{ ohms}$$

Decreasing the capacitance increases the reactance.

These examples show how the reactance is computed when the frequency and the capacitance are known. There are often occasions when it is necessary to compute the frequency when the reactance and capacitance are known or to compute the capacitance when the reactance and frequency are known. The formulas for making these calculations are given below:

$$f = \frac{159200}{X_cC}$$

$$C = \frac{159200}{fX_c}$$

$f =$ frequency in Hz
$C =$ capacitance in μf
$X_c =$ reactance in ohms

Examples:

1. At what frequency will a .039μf capacitor have a reactance of 7000 ohms?

$$f = \frac{159200}{7000(.039)} = 583.1 \text{ Hz}$$

2. What value of capacitance has a reactance of 23 k ohms at a frequency of 2 MHz.

$$23 \text{ k} = 23000 \text{ ohms}$$

$$2 \text{ MHz} = 2,000,000 \text{ Hz}$$

$$C = \frac{159200}{23000(2000000)} = .0000034\mu f$$

$$C = .0000034 \times 10^6 = 3.4\text{pf}$$

Ohm's Law in Capacitive Circuits

In an ac circuit, a capacitor is just as effective as a resistor in controlling the amount of current. Ohm's law describes the basic relationship between the current, voltage and resistance in an electrical circuit. The relationship holds true with capacitive as well as resistive ac circuits. In a purely capacitive circuit, one which has capacitance but no resistance, the current in the circuit is directly proportional to the applied voltage and inversely proportional to the capacitive reactance. This relationship is expressed in the formula below:

$$I = \frac{E}{X_c}$$

For example, assume that an ac voltage of 6.3 volts is applied to a capacitor with a reactance of 210 ohms. The current in the circuit then is:

$$I = \frac{E}{X_c} = \frac{6.3}{210} = .03 \text{ amperes or 30 milliamperes}$$

Keeping in mind that the capacitive reactance is a function of the frequency of the applied ac voltage and the capacitance, consider the effect that each of these factors has upon the current. Since the capacitive reactance is inversely proportional to the frequency and the capacitance, the current in the circuit will be directly proportional. Increasing the frequency or the capacitance causes the reactance to decrease. This decreases the opposition to current flow, therefore the current increases. Decreasing the frequency or capacitance causes the reactance to increase. This increases the opposition to current flow in the circuit and decreases the current.

Rearranging the basic Ohm's law formula given above, we can compute voltage and reactance.

$$E = IX_c$$

$$X_c = \frac{E}{I}$$

Some examples will illustrate the use of Ohm's law in these cases.

1. A .05 μf capacitor has a current of .1 ampere flowing through it at 400 Hz. What is the voltage across it?

$$X_c = \frac{159200}{fC} = \frac{159200}{400(.05)} = 7960 \text{ ohms}$$

$$E = IX_c = .1 \ (7960) = 796 \text{ volts}$$

2. A capacitor has 200 volts of 60 Hz ac across it and a current of .02 amperes through it. What is the value of capacitance?

$$X_c = \frac{E}{I} = \frac{200}{.02} = 10,000 \text{ ohms}$$

$$C = \frac{159200}{fX_c} = \frac{159200}{60(10000)} = .265 \mu f$$

RC CIRCUITS

While purely capacitive circuits are sometimes used in electronics, more often capacitors are combined with other components to form electronic networks. The most commonly used circuit is a series resistor and capacitor. Despite its simplicity, this simple series RC circuit has many applications. Another commonly used capacitor circuit is the parallel resistor-capacitor combination. While not as common as the series RC circuit, the parallel RC circuit is frequently found in electronic equipment. In this section you are going to investigate the operation and characteristics of both series and parallel RC circuits.

Series RC Circuits

The simplest form of RC circuit is a single capacitor connected in series with a resistor. Figure 3-17 shows such a circuit connected to a source of ac voltage. The source voltage generates a sine wave designated E. The applied voltage will cause current to flow in the circuit. The capacitor will charge and discharge as the ac voltage varies. No electrons pass through the capacitor, but the charging and discharging action causes a movement of electrons in the circuit. The charge on the capacitor causes a voltage to be developed across it. This voltage is designated E_C as shown in Figure 3-17. The voltage across the capacitor is a sine wave that lags the current flowing in the circuit by 90°. As indicated earlier, the current in a capacitive circuit leads the voltage across the capacitor by 90°.

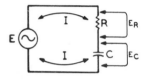

Figure 3-17
Series RC circuit.

The current in the circuit passes through resistor R and develops a voltage drop across it. The voltage across the resistor, designated E_R, is in phase with the current in the circuit. The voltage across the resistor is a function of the resistance and the current (E_R = IR). The voltage across the capacitor is determined in the same way. The capacitor voltage drop is a function of the current flowing in the circuit and the capacitive reactance or E_C = IX_C.

Figure 3-18 shows the phase relationships of the various voltages and currents in this series RC circuit. In this illustration, the current sine-wave (I) is used as the reference. As you recall, one of the fundamental characteristics of a series circuit is that the current is common to all components. Note the voltage drop across the resistor E_R. This voltage drop is in phase with the current. You can see this by noting that the maximum, minimum, and zero crossing points for the voltage waveform coincide with those of the current waveform.

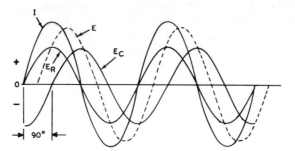

Figure 3-18 Phase relationships between the current and voltage in a series RC circuit.

Now refer to the capacitive voltage drop E_C. The current through a capacitor leads the voltage across it by 90°. Another way of saying this is that the voltage across the capacitor lags (occurs later in time) the current.

Another important characteristic of capacitive ac circuits is that Kirchhoff's law also applies. Kirchhoff's voltage law tells us that the sum of the voltage drops across the components in a series circuit equals the applied voltage. This law is valid for the circuit in Figure 3-17. Now refer to Figure 3-18. If we add E_R and E_C by summing the amplitudes of the two sine waves at multiple points and plotting the resulting curve, we will obtain the sine wave represented by the dashed line in Figure 3-18. This waveform represents the applied voltage E. Note that its amplitude is higher than either the resistor voltage or the capacitor voltage as you would expect. Another important point to note is that the applied voltage is not in phase with the current or with either the capacitor or resistor voltages. The current in the circuit leads the applied voltage as it will in any capacitive circuit. But since the circuit is not purely capacitive (because of the series resistor), the current will lead the voltage by some phase angle less than 90°. In this example, the phase shift between the current and the applied voltage is approximately 45°. In a purely capacitive circuit (no resistance), the current and the applied voltage will be out of phase by 90°. In a purely resistive circuit the current and voltage will be in phase. With both resistance and capacitance in the circuit, the phase difference between the applied voltage and the current will be some value between 0° and 90°. The exact amount of phase shift will be determined by the size of the resistance and the capacitive reactance.

When using Kirchhoff's law to analyze a series RC circuit, you should realize that it is not possible to directly add the numerical values of the resistor and capacitor voltages to obtain the applied voltage. In other words, the relationship below is incorrect:

$$E = E_R + E_C$$

In a purely resistive circuit where the current and voltage are in phase, the voltage drops across the various series elements in the circuit can be summed directly to obtain the applied voltage. This is not true in an ac circuit using capacitance because of the phase difference between the voltage across the capacitor and the voltage across the resistor. To obtain the correct sum, the phase angle between the resistor voltage and the capacitor voltage must be taken into consideration. This can be accomplished by vector or phasor addition.

Vector Diagrams. A vector or phasor is a line whose length represents the peak value of an ac voltage or current. The direction in which the vector points indicates its phase relationship to other vectors. Various vectors are combined to form a vector diagram which shows the phase and amplitude relationships of currents and voltages in a reactive ac circuit. A vector diagram can be drawn to represent the current and voltages in the series ac circuit we have been analyzing.

Figure 3-19 shows the basic format for drawing vector diagrams. Here a vector labeled I is drawn starting at the origin (designated O) in the center of the diagram. The point of the arrow is designated A. The distance between O and A represents the peak value of an alternating current (I). Note that the current phasor is pointing to the right which is designated as the 0° position. The phasor is assumed to be rotating in the counterclockwise direction. One complete 360° rotation corresponds to one cycle of the ac sine wave represented by the vector. The position of the vector anywhere during its rotation is representative of the particular point in the occurrence of one cycle of the sine wave. By adding other phasors to the diagram, a complete picture of the currents and voltages in the ac circuit can be obtained.

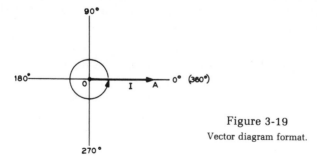

Figure 3-19
Vector diagram format.

127

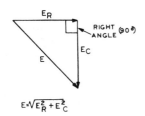

$$E = \sqrt{E_R^2 + E_C^2}$$

Figure 3-21 Voltage vector triangle for a series RC circuit

Figure 3-20 shows the phasor diagram for a series RC circuit. A current vector (I) is shown pointing to the right. This vector represents the peak value of the current flowing in the series RC circuit. Coinciding with the current vector is another vector labeled E_R. The length of the vector from the origin (O) to the point of the arrow represents the peak value of the voltage across the resistor. The resistive voltage vector overlaps or coincides with the current vector because these two signals are in phase.

The voltage across the capacitor E_C is 90° out of phase with the current. This is illustrated by the vector labelled E_C in Figure 3-20. Note that the direction of the capacitor voltage is shifted 90° from the direction of the current vector. The length of the vector represents the peak value of the capacitor voltage.

To find the applied voltage (E), we must add the capacitor and resistor voltages. Since they are out of phase with one another, the peak values cannot be added directly as indicated earlier. Instead, the addition must take into consideration both the magnitudes of the voltages and their directions.

To accomplish the addition of the resistor and capacitor voltages in graphical manner, a rectangle can be formed by using the capacitor and resistor voltage vectors as shown in Figure 3-20. The dashed lines in the diagram indicate how the rectangle is completed. The magnitude of the applied voltage is the distance between the origin and the far corner formed by completing the rectangle. In other words, the applied voltage in the circuit is represented by the diagonal line drawn from the origin. The angle formed by the applied voltage vector (E) and the current vector (I) represents the amount of phase shift in the circuit. This phase angle is designated theta (θ). Note that it is some value less than 90°.

There is a more direct way of computing the magnitude of the applied voltage if the resistor and capacitor voltages are known. In Figure 3-20 you found the applied voltage by graphical means. You completed the rectangle whose sides are equal to the capacitor and resistor voltages. Then you drew the diagonal line which represents the magnitude of the applied voltage. This graphical technique can be reduced to a simple mathematical formula for computing the applied voltage.

In Figure 3-20, consider the triangle formed by the resistor voltage E_R, the applied voltage E and the dashed line connecting these two vectors. The dashed line has the same length as the vector representing the capacitor voltage. You can use Pythagorean's theorem to solve this right triangle.

This theorem states a method of computing the length of one side of a right triangle if any of the other two sides are known. This is a basic geometric theorem that has many applications.

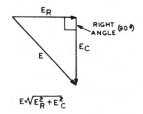

Refer to Figure 3-21. The vector representing the applied voltage is the side of the triangle directly opposite the 90° angle. This side is called the hypotenuse. The formula for computing the hypotenuse (applied voltage) is given below:

$$E = \sqrt{(E_R)^2 + (E_C)^2}$$

Figure 3-21 Voltage vector triangle for a series RC circuit

This expression says that if you square the resistor voltage and add it to the square of the capacitor voltage and take the square root of the sum, the result will be the applied voltage.

The expression above for finding the applied voltage when the resistor and capacitor voltages are known can be rearranged by simple algebra so that the resistor voltage can be computed if the applied voltage and capacitor voltage are known, or the capacitor voltage can be computed if the applied voltage and resistor voltage are known. These two formulas are given below:

$$E_R = \sqrt{(E)^2 - (E_C)^2}$$

$$E_C = \sqrt{(E)^2 - (E_R)^2}$$

The examples below show the use of the formulas.

1. What is the applied voltage in a series RC circuit where the resistor voltage is 12 volts and the capacitor voltage is 18 volts?

$$E = \sqrt{(E_R)^2 + (E_C)^2}$$

$$E = \sqrt{12^2 + 18^2}$$

$$E = \sqrt{144 + 324}$$

$$E = \sqrt{468}$$

$$E = 21.63 \text{ volts}$$

2. The voltage applied to a series RC circuit is 115 volts. The current in the circuit is .0075 amperes. The resistor value is 5 k ohms. What is the capacitor voltage?

$$5k = 5000$$

First, compute the resistor voltage:

$$E_R = IR$$

$$E_R = .0075 \ (5000)$$

$$E_R = 37.5 \ volts$$

Next compute the capacitor voltage:

$$E_C = \sqrt{(E)^2 - (E_R)^2}$$

$$E_C = \sqrt{115^2 - 37.5^2}$$

$$E_C = \sqrt{13225 - 1406.25}$$

$$E_C = \sqrt{11818.75}$$

$$E_C = 108.7 \ volts$$

Impedance

Impedance is the total opposition to current flow in an ac circuit. In a circuit consisting of a resistor and a capacitor, the total opposition is the sum of the capacitive reactance and the resistance. Both the reactance and the resistance impede current flow. Because of the phase shift created by the capacitor, the capacitive reactance and resistance cannot be summed directly, just as the voltage drops across the capacitor and resistor in a series RC circuit cannot be summed directly. The impedance is the vector sum of the capacitive reactance and resistance as you shall see.

The impedance of an ac circuit is expressed in ohms and is designated by the letter Z. We can define the impedance in terms of Ohm's law just as we defined the total resistance of a dc circuit. The impedance of an ac circuit is equal to the applied voltage divided by the total circuit current.

$$Z = \frac{E}{I}$$

This expression can be rearranged using basic algebra to obtain the expressions for voltage and current in terms of the circuit impedance:

$$E = IZ$$

$$I = \frac{E}{Z}$$

In the previous section, you saw that because of the phase shift caused by the capacitor in a series RC circuit, the voltage drops across the capacitor and resistor could not be added directly to obtain the applied voltage. Instead, a vector sum had to be taken in order to obtain the correct value. The voltage drops across the resistor and capacitor in a series RC circuit are directly proportional to the current. Since the current through a series circuit is the same in all elements, we can say that the voltage drops across the circuit components are directly proportional to their resistance or reactance. For that reason, we can draw a diagram exactly like the voltage vector diagram described earlier, to obtain the total impedance of the circuit. This vector diagram is shown in Figure 3-22. Here the current vector (I) is used as the reference as before. The resistance vector coincides with the current vector since the resistive voltage drop is in phase with the current. In this case, the length of the vector is proportional to the resistance. Another vector representing the magnitude of the capacitive reactance (X_c) is drawn 90° out of phase with the resistance vector to take into account the 90° phase shift produced by the capacitor. Completing the rectangle formed by the resistance and reactance vectors and drawing the diagonal line as indicated, gives us the magnitude of the impedance. The length of the diagonal line represents the total opposition to current flow. Using Pythagorean's theorem, we can write an expression for the impedance of the circuit in terms of the resistance and reactance:

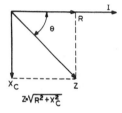

Figure 3-22 Impedance vector diagram of a series RC circuit

$$Z = \sqrt{R^2 + X_c^2}$$

This expression says that the impedance is equal to the square root of the sum of the resistance squared and the capacitive reactance squared. Using basic algebra, this formula can be rearranged to compute the circuit resistance in terms of the impedance and reactance or the capacitive reactance in terms of the impedance and resistance:

$$R = \sqrt{Z^2 - X_c^2}$$

$$X_c = \sqrt{Z^2 - R^2}$$

The examples below illustrate the use of these formulas.

1. What is the impedance of a series RC circuit with a resistance of 50 ohms and a reactance of 75 ohms?

$$Z = \sqrt{R^2 + X_c^2}$$

$$Z = \sqrt{50^2 + 75^2}$$

$$Z = \sqrt{2500 + 5625}$$

$$Z = \sqrt{8125}$$

$$Z = 90.14 \text{ ohms}$$

2. A series RC circuit has an applied voltage of 120 volts. The current is .02 amperes. The circuit resistance is 4000 ohms. What is the capacitive reactance?

$$Z = \frac{E}{I}$$

$$Z = \frac{120}{.02} = 6000 \text{ ohms}$$

$$X_c = \sqrt{Z^2 - R^2}$$

$$X_c = \sqrt{6000^2 - 4000^2}$$

$$X_c = \sqrt{36000000 - 16000000}$$

$$X_c = \sqrt{20000000}$$

$$X_c = 4472 \text{ ohms}$$

Phase Shift

In a capacitive ac circuit, the current will always lead the applied voltage. In a purely capacitive circuit, the phase shift between the current and the applied voltage will be 90°. When a series resistance is introduced into the circuit, the phase shift will be less than 90°. A purely resistive circuit, of course, will introduce no phase shift. The exact amount of phase shift in the circuit is determined by the size of the capacitor, the frequency of operation and the value of the resistance. In other words, the phase shift is a function of the reactance and the resistance. To be more specific, the phase shift is a function of the ratio of the capacitive reactance to the resistance : $\dfrac{X_c}{R}$.

The larger the capacitive reactance with respect to the resistance, the greater the phase shift. As the capacitive reactance increases or the resistance decreases, the phase shift will approach 90°. As the capacitive reactance decreases or the resistance increases, the phase shift will approach 0°.

In most applications, it is desirable to know the exact amount of phase shift that exists. In some cases the phase shift may be detrimental. Steps must usually be taken to correct for the phase shift or to make provisions for accommodating it. In other applications, the phase shift may have been introduced to produce a desired result. While phase shift can be measured with electronic instruments such as an oscilloscope, it can also be calculated directly if the values of resistance, capacitance and frequency of operation are known.

The phase shift in an RC circuit is computed by the use of trigonometry. Trigonometry is the science of dealing with triangles, and it provides methods for computing the sides and angles of any triangle. As you saw in the discussion of vector diagrams, the phase angle of an RC circuit is that angle formed between the resistor voltage and applied voltage or between the resistance vector and impedance vector. The lengths of these sides determine the size of the phase angle. By knowing the resistance and capacitive reactance or the capacitor voltage and the resistor voltage, the amount of phase shift in the circuit can be computed by the use of trigonometry.

The basic trigonometric expression for computing the phase shift in a series RC circuit is given below. The phase angle is θ (theta).

$$\theta = \arctan \frac{X_c}{R}$$

If the capacitive reactance and resistance are known, the phase angle can be found by computing the ratio of the reactance to the resistance. This gives you the tangent of the phase angle. The tangent can then be looked up in a set of trig tables (see Appendix A) to determine the phase angle. If the capacitor and resistor voltage in a series RC circuit are known, the phase angle can be found with the expression:

$$\theta = \arctan \frac{E_c}{E_R}$$

The formulas below illustrate how the phase shift, the resistance, the reactance and the impedance of a series RC circuit are related. Note that if any two of these four quantities are known the others can be computed with these expressions:

$$\sin \theta = \frac{X_c}{Z}$$

$$\cos \theta = \frac{R}{Z}$$

$$\tan \theta = \frac{X_c}{R}$$

$$R = Z \cos \theta$$

$$R = \frac{X_c}{\tan \theta}$$

$$X_c = Z \sin \theta$$

$$X_c = R \tan \theta$$

$$Z = \frac{X_c}{\sin \theta}$$

$$Z = \frac{R}{\cos \theta}$$

$$\theta = \arcsin \frac{X_c}{Z}$$

$$\theta = \arccos \frac{R}{Z}$$

$$\theta = \arctan \frac{X_c}{R}$$

The example problems below illustrate the use of these formulas. Trig tables are given in Appendix A.

1. What is the phase shift in a series RC circuit with a resistance of 1200, a current of 0.02 amperes and an applied voltage of 48 volts?

$$Z = \frac{E}{I} = \frac{48}{.02} = 2400 \text{ ohms}$$

$$\theta = \text{arccos} \frac{R}{Z} = \text{arccos} \frac{1200}{2400} = \text{arccos } .5$$

$$\theta = 60°$$

2. The measured phase angle in a series RC circuit is 40°. The circuit resistance is 2200 ohms. What is the capacitive reactance and impedance?

$$X_c = R \tan \theta = 2200 \tan 40°$$

$$2200 \, (.839) = 1845.8 \text{ ohms}$$

$$Z = \frac{R}{\cos \theta} = \frac{2200}{\cos 40°} = \frac{2200}{.766} = 2872 \text{ ohms}$$

Power in AC Circuits

Whenever a dc voltage source is connected to a resistor, current flows and power is dissipated in the form of heat. Power is a measure of the energy used in converting electrical energy into heat energy. The same effect occurs when an ac voltage is applied to a resistance. Current flows and electrical energy is converted into heat energy. The heat energy or power is dissipated in the resistance. All of the basic formulas for computing power dissipation in a dc circuit are applicable to power computation in an ac circuit:

$$P = EI$$

$$P = I^2R$$

$$P = \frac{E^2}{R}$$

When using these formulas for computing ac power dissipation, the voltage (E) or current (I) should be the effective or rms value.

Figure 3-23 shows a diagram of the current, voltage and power in a resistive ac circuit. Note that the applied voltage and current are in phase with one another. To obtain the power, we multiply the instantaneous voltage and current values and plot the resulting products. When a positive quantity (voltage) is multiplied by another positive quantity (current) the product will be positive. In addition, whenever two negative quantities are multiplied, the product is also positive. Note that the power curve in Figure 3-23 is all positive and sinusoidal in shape. The power waveform is also twice the frequency of the voltage and current waveforms. If we draw a line half-way between the maximum and zero points on the power curve, we will have a value that represents the average power dissipated by the resistance. This value of average power is the product of the rms values of current and voltage in the circuit.

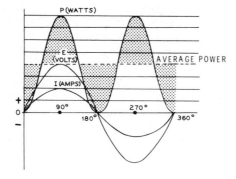

Figure 3-23

Power in a resistive ac circuit.

137

The most important point to remember here is that power is dissipated only in resistance. As you will see, power *appears* to be dissipated in reactive components such as capacitors and inductors. However, because of their unique nature and the phase shift they introduce between the current and voltage, they do not dissipate power. A rule to remember when computing the power in an ac circuit containing capacitors or inductors, is simply determine the circuit resistance and either the current through that resistance or voltage across it and use the previously given power formulas.

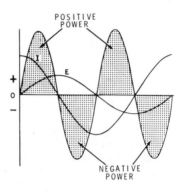

Figure 3-24

Power in a purely capacitive circuit.

Figure 3-24 shows what happens when an ac voltage sine wave is applied to a capacitor. The current in the circuit leads the applied voltage by 90°. To compute the power in the circuit, you multiply the instantaneous values of current and voltage and plot the resulting power curve. Unlike the resistive power curve in Figure 3-23, the power curve in a purely capacitive circuit is both positive and negative. What this power curve tells us is that during one-half cycle of the applied voltage, the capacitor appears to consume power. This is indicated by the positive part of the power curve. During the other half cycle, the power is negative. It is during this time that the capacitor actually acts as the supply and furnishes power to the source. When the capacitor charges it consumes power. When it discharges, it gives power back to the circuit. Since the positive and negative power curves are equal and opposite, the total average effect is zero. This means that no power is dissipated in purely capacitive circuit.

Figure 3-25 illustrates the power in an ac circuit containing both capacitance and resistance. Note that the current leads the voltage by some angle less than 90°. With this condition, the positive power is greater than the negative power. The negative power is still the result of the ability of a capacitor to store a charge and then return it to the circuit at another part of the cycle. The larger positive power in the circuit indicates that the resistor is dissipating power in the form of heat. A purely capacitive circuit dissipates no power. A purely resistive circuit dissipates all of the power in the form of heat. But in a circuit containing both resistance and reactance, power is dissipated only in the resistance.

Power Factor. If we use the formula P = EI to compute the power in a circuit containing both resistance and capacitance, we will obtain a value called the apparent power. We call it apparent power because power *appears* to be dissipated since an applied voltage is causing current to flow. In a purely capacitive circuit, current does flow when an ac voltage is applied. However no real power is dissipated. In a purely reactive circuit, one containing only a capacitor, this apparent power is sometimes referred to as reactive power. In a circuit containing both resistance and capacitance, the apparent power includes both the true power dissipated in the circuit resistance and the power that appears to be dissipated in the capacitor.

The power dissipated in a resistance is called true power. You can find the true power by knowing the resistance, the current through it or the voltage across it and using one of the power formulas given earlier. True power is always less than the apparent power.

The ratio of the true power to the apparent power in an ac circuit is referred to as power factor (PF). This is expressed in the formula below:

$$PF = \frac{\text{true power}}{\text{apparent power}}$$

The power factor is an excellent indication of the relative amounts of resistance and reactance in a given circuit. In a purely resistive circuit, the true power and the apparent power will be equal. Therefore the power factor will be equal to one. In a purely reactive circuit, such as one containing only a capacitor, the true power will be zero. Correspondingly, the power factor will be zero. For circuits containing both resistance and reactance, the power factor will be some value between zero and one. The greater the power factor, the more resistive the circuit. The lower the power factor, the greater the circuit reactance.

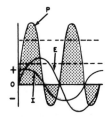

Figure 3-25
Power dissipation in a
series RC circuit.

In a series RC circuit, the power factor is determined by the values of the resistance and the reactance in the circuit. In fact, the power factor is equal to the ratio of the resistance to the total impedance of the series RC circuit:

$$PF = \frac{R}{Z}$$

The ratio of the resistance to the impedance is also equal to the cosine of the phase angle:

$$\cos \theta = \frac{R}{Z} = power\ factor$$

Parallel RC Circuits

Another often encountered RC circuit is the parallel combination of a resistance and capacitor. See Figure 3-26. Here the applied ac voltage is common to both the resistor and the capacitor. Current flowing in the resistor is determined by the applied voltage and the resistance:

$$I_R = \frac{E}{R}$$

The current in the capacitor is a function of the applied voltage and the capacitive reactance:

$$I_C = \frac{E}{X_c}$$

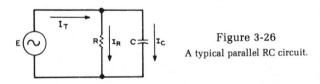

Figure 3-26
A typical parallel RC circuit.

The total current drawn from the ac source is the sum of resistor and capacitor currents. The current through the resistor is in phase with the applied voltage, but the current through the capacitor leads the applied voltage by 90°. Because of this phase relationship, the currents cannot be added directly to determine total current drawn from the source. The total current then is the vector sum of the resistor and capacitor currents.

This is illustrated by the vector diagram shown in Figure 3-27. In this vector diagram the applied voltage E is used as the reference vector since it is common to both circuit elements. The resistor current I_R is represented by the vector coinciding with the applied voltage vector. The resistor current is in phase with the applied voltage. The capacitor current (I_C) is 90° out of phase with the applied voltage. Note the direction of I_C in Figure 3-27. All vectors are assumed to be rotating in a counterclockwise direction. Therefore, to show the capacitor current leading the applied voltage, it is drawn shifted from the applied voltage vector by 90°. The total circuit current (I_T) then can be obtained by vector addition of the resistor and capacitor currents. Using Pythagorean's theorem, the expression is shown below:

$$I_T = \sqrt{(I_R)^2 + (I_C)^2}$$

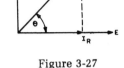

Figure 3-27

Current vector diagram of a parallel RC circuit.

Note in Figure 3-27 that the total current drawn from the ac source leads the applied voltage. Since the circuit is capacitive, the current will always lead the applied voltage. However, since the circuit is not purely capacitive, the current leads by some phase angle less than 90°.

The impedance of a parallel RC circuit can be expressed in terms of Ohm's law. The impedance of a parallel RC network is the applied voltage divided by the total current:

$$Z = \frac{E}{I_T}$$

The impedance of a parallel RC circuit can also be found by using the general formula for finding the total resistance of two resistors connected in parallel. This formula is indicated below:

$$R_T = \frac{R_1 R_2}{R_1 + R_2}$$

If the resistance and capacitive reactance are known, the impedance can be found with this expression:

$$Z = \frac{RX_C}{\sqrt{R^2 + X_c^2}}$$

APPLICATIONS OF CAPACITIVE CIRCUITS

In this section you will learn some of the most important applications of capacitors in ac circuits. While capacitors are often used alone, usually they are combined with resistors or other components to form RC networks. Such networks have many practical applications. These include ac voltage dividers, filters, decouplers, phase shifters and coupling networks. Each of these important applications is discussed here.

Capacitive Voltage Dividers

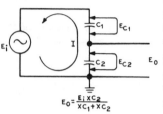

$$E_o = \frac{E_i \, x_{C_2}}{x_{C_1} + x_{C_2}}$$

Figure 3-28
A capacitive voltage divider.

A capacitive voltage divider is a series capacitive circuit whose output is some fraction of its input. Figure 3-28 shows a simple capacitive voltage divider made by connecting two capacitors, C_1 and C_2 in series. The input voltage (E_i) is connected to both capacitors. The output voltage (E_0) is taken from across capacitor C_2. The current flowing in the circuit produces a voltage drop across each capacitor. The sum of the capacitive voltage drops is equal to the input voltage as the expression below indicates.

$$E_i = \; E_{C1} + E_{C2}$$

In this application, the output voltage is equal to the voltage across the capacitor C_2.

$$E_o = \; E_{C2}$$

The amount of voltage developed across each capacitor is a function of a current in the circuit and the capacitive reactance according to Ohm's law. The output voltage will then be:

$$E_o = \; E_{C2} = IX_{C2}$$

The voltage drops across the capacitors in the circuit, divide in proportion to their capacitive reactances. The greater the capacitive reactance, the greater the voltage drop across that capacitor. Keep in mind that the reactance is a function of the size of the capacitor and the frequency of the applied voltage.

The output voltage can be expressed in terms of the input voltage and the capacitive reactances, as shown below:

$$E_o = \; \frac{X_{C2}}{X_{C1} + X_{C2}} \; (E_i)$$

142

As you can see from this expression, the output voltage is equal to the input voltage multiplied by the ratio of the reactance of C_2 to the sum of the individual reactances. This ratio is referred to as the voltage division ratio.

We can also express the output voltage in terms of the capacitor values rather than the reactances. The formula below gives this relationship. Note that the output voltage is directly proportional to the voltage division ratio which is the ratio of C_1 to the sum of C_1 and C_2.

$$E_o = \frac{C_1}{C_1 + C_2} (E_i)$$

An interesting characteristic of a capacitive voltage divider is that the voltage division ratio is not affected by frequency. Even though the frequency causes the reactances to change, the reactances change together and the voltage division ratio remains constant.

Another interesting characteristic of the purely capacitive voltage divider is that no phase shift occurs between the input and the output. The current in this circuit leads the applied voltage, but the voltage across the output capacitor is in phase with the input voltage. Again, changing the frequency has no effect on this characteristic.

Capacitive voltage dividers are often found in high frequency amplifier circuits. Certain types of oscillators also use capacitive voltage dividers. The output voltage from a capacitive voltage divider can be made variable by making either C_1 or C_2 a variable capacitor. Changing the value of capacitance will permit the output voltage to be adjusted to a desired value.

RC Filters

A filter is a frequency discriminating circuit. Filters greatly attenuate some frequencies while allowing others to pass with virtually no opposition. Filters are frequency selective in that they pass some frequencies and attempt to keep others out.

Two of the most common types of filters used in electronic circuits are the low-pass filter and the hi-pass filter. A low-pass filter permits low frequency signals to pass from the input to the output with little or no attenuation. High frequencies, however, are greatly attenuated. A frequency known as the cut-off frequancy (f_{co}) is the general dividing line between those frequencies that are passed and those that are attenuated.

A high-pass filter is the opposite of a low-pass filter. The high-pass filter permits frequencies above the cut-off frequency to pass. Frequencies below the cut-off point are greatly attenuated.

Simple RC networks can be used as low and high-pass filters. Such circuits are able to perform a frequency selective function because of the change in reactance with frequency.

Low-Pass Filter. The simplest form of low-pass filter is shown in Figure 3-29A. It consists of a resistor and capacitor connected in series across an input voltage (E_i). The output voltage is taken from across the capacitor. Assume that the input voltage E_i has a fixed rms value but that its frequency can be varied.

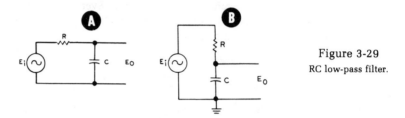

Figure 3-29

RC low-pass filter.

The best way to understand the operation of the low-pass filter is to look at the circuit as a voltage divider. See Figure 3-29B. The input voltage is applied across the resistor and capacitor in series. The output voltage is taken from across the capacitor. The voltage division ratio depends upon the sizes of the resistance and the capacitive reactance. The value of the resistance remains constant, of course, but the value of the capacitive reactance changes as the input frequency changes.

At very low input frequencies, the capacitive reactance will be very high. If the reactance is high compared to the resistance, most of the input voltage will appear across the capacitor. At low frequencies then, the circuit offers very little opposition, and nearly all the input voltage appears at the output. As the input frequency increases, the capacitive reactance decreases. This means that less voltage will be dropped across the capacitor and more across the resistor as the frequency gets higher. For this reason, the output voltage begins to drop off as frequency is increased. At very high frequencies, the reactance will be very low. If it is significantly lower than the value of the resistance, then very little voltage will appear at the output.

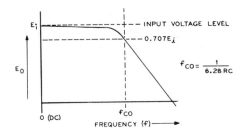

Figure 3-30 Frequency response
of an RC low-pass filter.

The frequency response curve shown in Figure 3-30 illustrates this effect. This curve shows the amount of output voltage (E_o) with respect to frequency (f). On the left-hand side of the curve, at very low frequencies, the output voltage is nearly equal to the input voltage. In fact, with a frequency of 0 Hz or dc, the capacitor offers maximum opposition and the output voltage will be equal to the input voltage. As the frequency increase, the capacitive reactance begins to decrease. The output voltage then begins to drop off. After (to the right of) the cut-off frequency (f_{co}), the output voltage drops off at a constant rate. At the cut-off frequency, the output voltage is equal to approximately 70% of the input voltage or E_o = .707 E. The cut-off frequency is a function of the resistor and the capacitor values and is determined by the expression below:

$$f_{co} = \frac{1}{2\pi RC} = \frac{1}{6.28\ RC}$$

where R is in ohms and C is in farads.

The formula can be simplified by solving for 1/6.28 and expressing C in μf:

$$f_{co} = \frac{159200}{RC}$$

For example, the cut-off frequency of a circuit with a 10 k ohm resistor and .01μf capacitor is:

$$R = \quad 10 \text{ k ohm} = 10,000$$
$$C = \quad .01\mu f$$

$$f_{co} = \frac{159200}{10000(.01)} = 1592 \text{ Hz}$$

145

An important thing to note about an RC low-pass filter is that while the circuit is frequency selective the selectivity is very gradual. The output is not sharply defined at the cut-off frequency. Above the cut-off frequency, the higher frequencies are only attenuated, not cut out completely. In other words, the low-pass filter does pass frequencies higher than the cut-off frequency but they are more greatly attenuated than those frequencies below the cut-off point. Despite this imperfection in RC low-pass filters, these circuits are still very useful.

High-Pass Filter. A simple RC high-pass filter is shown in Figure 3-31A. Like the low-pass filter, it consists of a resistor and a capacitor connected in series to the input voltage. In the high-pass filter, however, the output voltage is taken from across the resistor. In Figure 3-31B, the circuit is drawn as a voltage divider.

At very high input frequencies, the capacitive reactance will be very low. If it is low compared to the resistance, little voltage will be dropped across it. At high frequencies then, most of the input voltage will appear across the resistor. As the frequency decreases, the capacitive reactance increases. More and more voltage will be dropped across the capacitance and less across the resistance. Therefore, as the frequency decreases, the output voltage will begin to decrease. The decrease is gradual at first, but at the cut-off frequency, the attenuation becomes more pronounced and the output voltage drops at a constant rate with decreasing frequency.

Figure 3-32 shows the frequency response curve of an RC high-pass filter. Note that at high frequencies the output voltage is nearly equal to the input voltage (E_i). As the frequency decreases, the output voltage begins to decrease. At the cut-off frequency, the output voltage is approximately 70% of the input voltage. Below the cut-off frequency, the attenuation increases and the output voltage drops.

As in the low-pass filter, the cut-off frequency is a function of the resistor and capacitor values. The same expression used for computing the cut-off frequency of a low-pass filter applies to the high-pass filter.

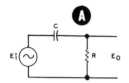

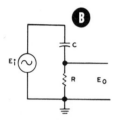

Figure 3-31
RC high-pass filter.

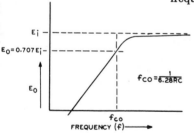

Figure 3-32 Frequency response of an RC high-pass filter.

$$f_{co} = \frac{159200}{RC}$$

$$R = \text{ohms}$$

$$C = \mu f$$

$$f_{co} = \text{Hz}$$

146

Circuits Combining AC and DC

Many electronic circuits operate from a combination of both ac and dc voltages. To these circuits, the applied voltage appears to be a dc voltage such as a battery connected in a series with an ac generator. The result is a dc level on which is superimposed an ac signal. We say that the ac rides on the dc. The zero line of the sine-wave signal coincides with the dc voltage level. RC networks are commonly used with such ac/dc source voltages.

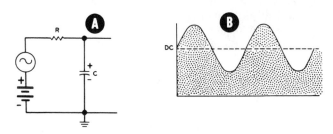

Figure 3-33
An RC decoupling network (A)
and the output waveform (B).

Decoupling Network. Figure 3-33A shows one application of an RC circuit with an ac/dc source. The RC circuit is connected as a low-pass filter. The capacitor will charge to the value of the dc voltage as indicated by the polarity shown. The sine-wave voltage will cause the charge on the capacitor to vary above and below the dc value by an amount equal to the peak value of the ac. Figure 3-33B shows the voltage across the capacitor. The combination voltage is often called pulsating dc. The composite voltage is still dc because it never goes negative, but it does vary or pulsate. In this example, the frequency of the ac voltage is assumed to be lower than the cut-off frequency of the RC low-pass network. If the frequency of the ac is higher than the cut-off frequency of the RC network, its amplitude will be greatly attenuated.

One of the most common applications for the low-pass RC network shown in Figure 3-33A is decoupling. Decoupling refers to a process of allowing a desired dc voltage to appear between given points while at the same time eliminating or minimizing the ac at that point. In many electronic circuits, the dc voltage is used to operate the equipment. AC signals in the form of oscillations, noise and transients sometimes appear as though they are connected in series with the dc supply voltage. The RC low-pass filter is used to eliminate the ac. By making the cut-off frequency of the RC network low enough, most or all of the undesired ac signals are filtered out leaving only the desired dc across the capacitor.

147

Coupling Networks. Another application of an RC circuit with ac/dc is shown in Figure 3-34. The source voltage is a combination of both ac and dc signals. This source is applied across a series RC network. However, in this application, the output is taken from across the resistor. In other words, the RC network is used as a high-pass filter. The capacitor charges to the applied dc voltage. Once it charges, however, no further dc flows in the circuit. Therefore, no dc appears across the output resistance. Instead, only the ac voltage will appear across the output. The sine-wave source will cause the capacitor to charge and discharge at the ac rate, thereby creating current flow in the resistor which appears at the output in the form of an ac voltage. In this application, only the ac signal appears at the output. The cut-off frequency of the RC network is adjusted so that it is low enough to permit the ac input to pass without noticeable attenuation.

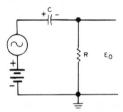

Figure 3-34 An RC coupling network.

This circuit is known as a coupling network. It is widely used to couple an ac signal from one point to another while blocking any dc. The capacitor prevents the dc from flowing in the output. Care should be taken in the selection of the resistor and capacitor values. Because of the high-pass filter effect they offer, their values should be chosen so that the cut-off frequency permits passage of the desired ac signal with little or no attenuation.

Phase Shift Networks

Another application for RC networks is phase shifting. Phase shifting is the process of shifting the phase of an output sine-wave signal with respect to the input. The process of phase shifting is sometimes used to correct undesirable phase shift in a circuit. In other applications, the phase shifting is done purposely to produce a desired effect. Since a capacitor causes the current in a circuit to lead the applied voltage, RC networks can be used in phase shifting applications.

148

Simple series RC networks are widely used in phase shift applications. Figure 3-35 shows the two most commonly used configurations. In Figure 3-35A, the input signal is applied across the RC combination and the output is taken from across the resistor. The current in the circuit leads the applied voltage by some phase angle between zero and 90° depending upon the resistor and capacitor values. The voltage across the resistor will be in phase with the current causing it. Therefore, the output voltage will lead the input voltage. This circuit produces a leading phase shift. The phase relationship between input and output is shown in Figure 3-36A.

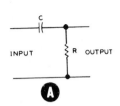

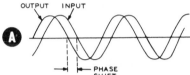

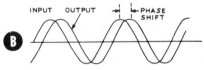

Figure 3-36

Phase shift in a series RC circuit (A) leading output (B) lagging output.

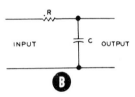

Figure 3-35

Basic RC phase shift networks (A) leading output (B) lagging output.

In the phase shift circuit of Figure 3-35B, the output voltage is taken from across the capacitor. Again the current in the circuit leads the applied voltage by some phase angle between zero and 90°. However, the voltage across the capacitor lags the applied voltage.

These simple RC phase shift networks are used where only small amounts of phase shift are required. When phase shifts of greater than approximately 60° are required, other phase shifting techniques are often used. The reason for this is that as the phase shift approaches 90° the output voltage drops to a very low level. Remember that these RC networks are also voltage dividers. Theoretically at 90°, the output voltage is zero.

Recall that the amount of phase shift in a simple RC circuit is a function of the reactance and the resistance in the circuit. The phase angle in the circuit is determined by the expression:

$$\theta = \arctan \frac{X_c}{R}$$

149

In order to approach a phase shift of 90° with a simple RC network, the ratio of the reactance to resistance will have to be extremely high. In the circuit of Figure 3-35A, most of the applied voltage would appear across the capacitor and very little would appear across the resistance. The output would be too low for most applications.

In the network of Figure 3-35B, the phase shift is:

$$\theta = \arctan \frac{R}{X_c}$$

Note that we use the ratio $\dfrac{R}{X_c}$ instead of $\dfrac{X_c}{R}$ as with the circuit of Figure 3-35A.

In the circuit of Figure 3-35B, the phase shift will approach 90° as the ratio of R/X_c increases. With this condition, most of the voltage will appear across the resistance and little across the capacitor. The phase shift will approach 90°. Again the output voltage would be too low to be usable for most applications requiring phase shifts near 90°.

When phase shifts of greater than approximately 60° are required, other phase shift techniques are usually used. One way to get greater phase shifts is to cascade the simple RC network shown in Figure 3-35. The output of one network can be connected to the input of another, thereby creating a total phase shift equal to the sum of the individual phase shifts. For example, if three 45° phase shift networks are cascaded as shown in Figure 3-37, the total phase shift would be $3 \times 45° = 135°$. In this case, the output voltage would lead the input voltage by 135°.

To achieve a lagging phase shift greater than that obtainable with a single RC network, the circuit of Figure 3-37B can be used. Here two networks are cascaded. For example, if each network provided a phase shift of 45°, the output would lag the input by a total of 90°.

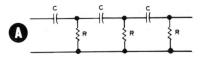

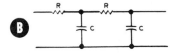

Figure 3-37
Cascade RC phase shift networks.

While cascaded RC networks can provide any degree of phase shift desired, keep in mind that each RC section acts as a voltage divider. This means that the output voltage will be much lower than the input voltage. Where such networks are used, amplifiers are generally required to offset the attenuation and bring the output voltage back up to a usable level.

Finally, it is important to note that the phase shift produced by an RC network is a function of the resistor and capacitor values. This phase shift is valid for only a single frequency. If an RC network is designed to produce a desired phase shift for one frequency, the amount of phase shift produced by that network at another frequency will be different. The reason for this is simple. If the frequency changes, the capacitive reactance also changes. Changing the reactance changes the phase angle.

Appendix A

TABLE OF TRIGONOMETRIC FUNCTIONS

Degrees	Sine	Cosine	Tangent	Degrees	Sine	Cosine	Tangent
0	0.000	1.000	0.000				
1	0.017	1.000	0.017	46	0.719	0.695	1.036
2	0.035	0.999	0.035	47	0.731	0.682	1.072
3	0.052	0.999	0.052	48	0.743	0.669	1.111
4	0.070	0.998	0.070	49	0.755	0.656	1.150
5	0.087	0.996	0.087	50	0.766	0.643	1.192
6	0.105	0.995	0.105	51	0.777	0.629	1.235
7	0.122	0.993	0.123	52	0.788	0.616	1.280
8	0.139	0.990	0.141	53	0.799	0.602	1.327
9	0.156	0.988	0.158	54	0.809	0.588	1.376
10	0.174	0.985	0.176	55	0.819	0.574	1.428
11	0.191	0.982	0.194	56	0.829	0.559	1.483
12	0.208	0.978	0.213	57	0.839	0.545	1.540
13	0.225	0.974	0.231	58	0.848	0.530	1.600
14	0.242	0.970	0.249	59	0.857	0.515	1.664
15	0.259	0.966	0.268	60	0.866	0.500	1.732
16	0.276	0.961	0.287	61	0.875	0.485	1.804
17	0.292	0.956	0.306	62	0.883	0.469	1.881
18	0.309	0.951	0.325	63	0.891	0.454	1.963
19	0.326	0.946	0.344	64	0.899	0.438	2.050
20	0.342	0.940	0.364	65	0.906	0.423	2.145
21	0.358	0.934	0.384	66	0.914	0.407	2.246
22	0.375	0.927	0.404	67	0.921	0.391	2.356
23	0.391	0.921	0.424	68	0.927	0.375	2.475
24	0.407	0.914	0.445	69	0.934	0.358	2.605
25	0.423	0.906	0.466	70	0.940	0.342	2.748
26	0.438	0.899	0.488	71	0.946	0.326	2.904
27	0.454	0.891	0.510	72	0.951	0.309	3.078
28	0.469	0.883	0.532	73	0.956	0.292	3.271
29	0.485	0.875	0.554	74	0.961	0.276	3.487
30	0.500	0.866	0.577	75	0.966	0.259	3.732
31	0.515	0.857	0.601	76	0.970	0.242	4.011
32	0.530	0.848	0.625	77	0.974	0.225	4.332
33	0.545	0.839	0.649	78	0.978	0.208	4.705
34	0.559	0.829	0.675	79	0.982	0.191	5.145
35	0.574	0.819	0.700	80	0.985	0.174	5.671
36	0.588	0.809	0.727	81	0.988	0.156	6.314
37	0.602	0.799	0.754	82	0.990	0.139	7.115
38	0.616	0.788	0.781	83	0.993	0.122	8.144
39	0.629	0.777	0.810	84	0.995	0.105	9.514
40	0.643	0.766	0.839	85	0.996	0.087	11.43
41	0.656	0.755	0.869	86	0.998	0.070	14.30
42	0.669	0.743	0.900	87	0.999	0.052	19.08
43	0.682	0.731	0.933	88	0.999	0.035	28.61
44	0.695	0.719	0.966	89	1.000	0.017	57.29
45	0.707	0.707	1.000	90	1.000	0.000	

Unit 4

INDUCTIVE CIRCUITS

INTRODUCTION

In this unit you are going to study inductors and how they are used in ac circuits. An inductor is an electronic component that opposes changes in current flow through it. Like a capacitor, it offers opposition to the flow of alternating current. When used in sinusoidal circuits, an inductor introduces a phase shift. Inductors can be combined with resistors and/or capacitors to form a variety of electronic circuits.

Your previous studies in dc electronics should have included inductors and inductive circuits. You should know what an inductor is, how it works and what effect it has in dc circuits. The first part of this unit is devoted to a review of inductors and inductance.

In the remaining sections, you will study the effects of inductance in an ac circuit. The characteristics and applications of simple resistor-inductor (RL) networks will also be considered.

Inductance is a characteristic found in virtually all electronic circuits. You must understand its effect in order to analyze, design or troubleshoot electronic circuits. In addition, your understanding of inductance is vital to your successful completion of the units on transformers and tuned circuits following this unit.

REVIEW OF INDUCTORS AND INDUCTANCE

When current flows through an electrical conductor such as a wire, a magnetic field will be generated around it. Each electron in the conductor has a minute magnetic field associated with it. The electrons in the conductor are aligned in a random manner when no current is flowing through it. For that reason, the tiny magnetic fields associated with each electron tend to cancel one another. With no current flowing through the conductor, no magnetic field exists. When a voltage is applied to the conductor, the electrons begin to flow. This tends to align the electrons so that their magnetic fields add. The total strength of the magnetic field is the sum of the individual electron fields. The higher the applied voltage and the lower resistance of the conductor, the greater the number of electrons that will flow. The magnitude of the magnetic field surrounding the conductor will increase as the amount of current flow increases. This effect is known as electromagnetism. See Figure 4-1.

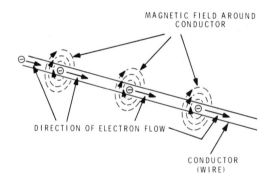

Figure 4-1
Magnetism around
a current-carrying conductor.

155

Current flowing in a conductor produces a magnetic field. In addition, a magnetic field can cause a current to flow in a conductor when there is relative motion between the magnetic field and the conductor. If a conductor or wire is passed through a stationary magnetic field, a voltage or electromotive force (emf) will be induced into that conductor. Alternately, if a magnetic field is passed across a fixed conductor, a voltage will be induced into the conductor. As long as there is relative motion between the magnetic field and the conductor, an induced voltage will be generated in the conductor. As the conductor is moved, the magnetic fields of the electrons in the conductor are affected by the external magnetic field. The motion between the conductor and the magnetic field forces the electrons to move in one direction or the other. The effect is that of creating a small voltage across the conductor where one end is more negative than the other. If the conductor forms a complete electrical circuit, current will flow in the circuit. See Figure 4-2.

The amount of voltage induced into a conductor by its motion in a magnetic field depends upon the strength of the magnetic field. The stronger the field, the more influence it will have on the electrons in the conductor and therefore the greater the induced voltage. The length of the conductor also determines the amount of induced voltage. For a given magnetic field, the longer the conductor, the greater the induced voltage. The speed with which the conductor cuts the magnetic field also influences the magnitude of the induced voltage. Moving the conductor through the field slowly causes only a small voltage to be induced. If the conductor is moved quickly through the magnetic field, a larger voltage is induced.

Figure 4-2
Electromagnetic
induction.

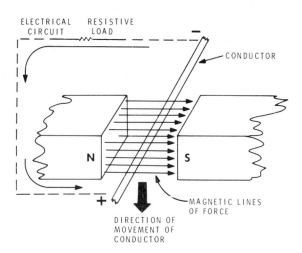

The direction of motion between the conductor and the magnetic field also determines the amount of induced voltage. If the conductor moves in the same direction as the magnetic field, no voltage will be induced. The conductor must cut across the magnetic lines of force in order for a voltage to be generated. If the conductor moves perpendicular to the magnetic lines of force, the maximum voltage will be induced. As you can see, the amount of induced voltage is directly proportional to the strength of the magnetic field, the length of the conductor, the speed of the conductor and the direction of movement of the conductor relative to the magnetic field. This effect is known as electromagnetic induction. Electromagnetism and electromagnetic induction are both responsible for the property of inductance and the effect it has on electrical circuits.

Self-Induction

When a voltage is applied to a conductor, current will flow. A magnetic field will be generated by the current flow. The moment that the voltage is applied to the conductor, no electrons flow and therefore no magnetic field will be developed. However, the instant that electrons begin to flow, the magnetic field begins to increase. The strength of the magnetic field builds up as the current flow rises from zero to its maximum value. As the magnetic lines of force expand outward from the center of the conductor, the magnetic field causes a voltage to be induced into the conductor itself. In other words, the magnetic field generated by the conductor causes a voltage to be induced into the conductor that produces the magnetic field. The expansion of the magnetic lines of force with respect to the conductor represents the relative motion between the conductor and the magnetic field required to induce a voltage. The polarity of the induced voltage opposes the polarity of the voltage applied to the conductor.

As long as the magnetic field is moving with respect to the conductor, the induced voltage will continue to be generated. As the magnetic lines of force continue to expand outward from the conductor during the rise of the current in the circuit, the induced voltage will be present. When the current in the circuit reaches its maximum level as determined by the amount of applied voltage and the resistance of the conductor, the magnetic field will become stationary. Since there is no further relative motion between the conductor and the magnetic field, there will be no induced voltage. At this time, the current in the circuit is strictly a function of Ohm's law.

If the voltage applied to the conductor is removed, current flow will cease. With no movement of electrons in the conductor, the magnetic field collapses. As it collapses, the lines of force cut across the conductor and induce a voltage. Again, the collapsing lines of force cause relative motion between the conductor and the magnetic field. Therefore, a voltage is induced in the conductor. The polarity of the induced voltage is such that it tends to keep current flowing in the same direction dictated by the external applied voltage.

The application or removal of the voltage source causes a self-induced voltage. This self induction takes place for any current changes that occur. Increasing or decreasing the current in a circuit causes the magnetic lines of force to expand or collapse and thereby cut the conductor and induce a voltage that will oppose the applied voltage. The induced voltage is referred to as counter emf or back emf, since it always opposes the applied voltage. The ability of a conductor to generate a voltage with a change in current is called self-induction. It is this characteristic that produces inductance.

Inductors and Inductance

Inductance is the property of an electrical circuit that tends to oppose any change of current in the circuit. The conductor or wire we have been discussing exhibits the property of inductance because it opposes changes in the current flow. If the current through a conductor suddenly increases, a voltage will be induced in the conductor that opposes the applied voltage. The induced voltage will attempt to cancel the applied voltage and will tend to hold the current to its previous level. The induced voltage opposes the applied voltage and therefore holds back the increase in the current. The current will still rise, however, because the induced voltage will appear only during the time that the current is increasing. Once there is no longer a relative motion between the conductor and the magnetic field, no further induced voltage or opposition takes place.

If the current in a circuit is suddenly decreased, the magnetic lines of force will contract or collapse and in doing so will induce a voltage in the conductor that opposes the applied voltage. If the applied voltage is suddenly decreased, the induced voltage will be in such a direction that it will tend to maintain the current at its previous higher level. The current will eventually decrease, however, as the magnetic field stops its collapse and no further induced voltage is generated.

As you can see, changing the current in a circuit causes a voltage to be induced in the conductor that opposes the change of current. The induced voltage tends to cause current in the circuit to remain at its previous level. This ability to oppose the change in an electrical current is referred to as inductance.

The electronic component that exhibits the property of inductance is called an inductor. The conductor or wire that we have been considering up to this point can be referred to as an inductor. While any wire or electrical conductor exhibits the property of inductance, it is normally not referred to as an inductor. Instead, an inductor is considered to be an electronic component. The most widely known inductor is a coil of wire. The term coil is often used interchangeable with the name inductor.

Whenever a wire is wound into a coil, the inductor thus formed becomes more manageable. Winding the wire into the coil makes the inductor smaller and more compact. At the same time, the inductance is greatly increased. By keeping the turns of wire close together, the magnetic field surrounding the wire will become more concentrated. The greater the magnetic field, the greater the induced voltage and therefore the higher the inductance.

Unit of Inductance

The unit of electrical inductance is the henry. One henry is defined as the amount of inductance that a coil has if the current, changing at the rate of one ampere per second, produces one volt of induced voltage. Inductance is a measure of how much counter emf is generated in an inductor for a specific amount of change in the current through that inductor.

The henry (abbreviated h) is a fairly large unit of inductance. While there are inductors available with an inductance of one henry or more, most inductors used in electronic circuits have a much lower inductance value. These inductance values are expressed in smaller units known as the millihenry (mh) and microhenry (μh). One millihenry is one thousandth of a henry (1 mh = 1/1000 h). One microhenry is one millionth of a henry (1μh = 1/1000000 h). One microhenry is also one thousandth of a millihenry (1μh = 1/1000 mh). The unit of inductance is usually expressed by the letter L.

Table I shows the relationship between the henry, the millihenry and the microhenry. Table II gives the rules for converting from one unit of inductance to another.

TABLE I
UNITS OF INDUCTANCE

1 henry = 1000 or 10^3 millihenry
1 henry = 1,000,000 or 10^6 microhenry
1 millihenry = 1/1000 or 10^{-3} henry
1 millihenry = 1000 or 10^3 microhenry
1 microhenry = 1/1000000 or 10^{-6} henry
1 microhenry = 1/1000 or 10^{-3} millihenry

Henry = h
Millihenry = mh
Microhenry = μh

TABLE II

Converting Units of Inductance

To Convert	To	Action
Henry	Millihenry	Multiply by 1000 (10^3) or move the decimal point three places to the right.
Henry	Microhenry	Multiply by 1,000,000 (10^6) or move the decimal point six places to the right.
Millihenry	Henry	Divide by 1000 (10^3) or move the decimal point three places to the left.
Millihenry	Microhenry	Multiply by 1000 (10^3) or move the decimal point three places to the right.
Microhenry	Henry	Divide by 1,000,000 (10^6) or move the decimal point six places to the left.
Microhenry	Millihenry	Divide by 1000 (10^3) or move the decimal point three places to the left.

The examples below show how the information in Tables I and II can be used to convert from one unit to another.

1. Convert 100 mh to h. 100 mh = 100 ÷ 1000 = .1h
2. Convert 78.6 μh to h. 78.6μh = 78.6 ÷ 1,000,000 = .0000786 h or 78.6 × 10^{-6} h
3. Convert 22 mh to μh. 22 mh = 22 × 1000 = 22000μh

Factors Affecting Inductance

It is the physical characteristics of a coil that determines its inductance. The amount of inductance that a coil has depends upon the number of turns in the coil, the spacing between the turns, the wire size, the shape of the coil, the number of layers of windings, the type of windings, the diameter of the coil, the length of the coil and the type of core material. Refer to Figure 4-3.

The inductance of a coil is directly proportional to the number of turns of wire and the diameter of the coil. The greater the number of turns the greater the magnetic field produced by the coil. In addition, for a given number of turns, the greater the diameter of the coil, the more magnetic lines of force produced. Anything done to increase the magnetic field produced by the coil will increase its inductance.

The spacing between the turns also affects the inductance. Keeping the turns very close together causes the magnetic lines of force produced by each turn to add together and produce a greater magnetic field. Spacing out the turns of the coil tends to reduce the addition of the lines of force produced by each turn, thereby decreasing the magnetic field. The spacing of the turns affects the coil length. The inductance is inversely proportional to the length. That is, increasing the length decreases the inductance.

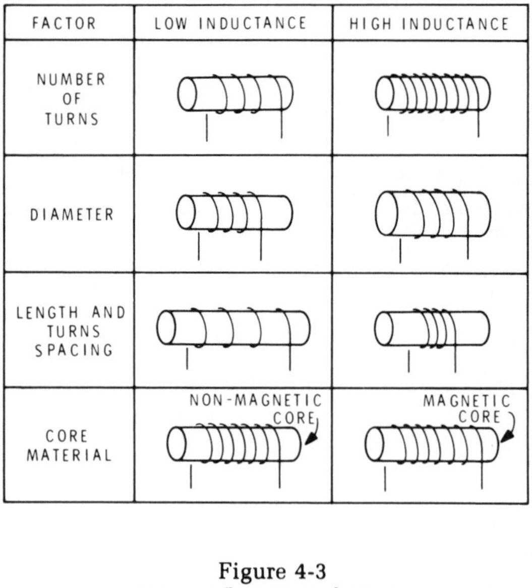

FACTOR	LOW INDUCTANCE	HIGH INDUCTANCE
NUMBER OF TURNS		
DIAMETER		
LENGTH AND TURNS SPACING		
CORE MATERIAL	NON-MAGNETIC CORE	MAGNETIC CORE

Figure 4-3
Factors influencing inductance.

162

The wire size used in the coil does not directly affect the inductance of the coil, however, it will influence the spacing between the turns when the turns are directly adjacent to one another. The smaller the wire, the closer the turns and the greater the inductance.

The number of layers of winding used will also affect the inductance. The greater the number of windings, the greater the inductance. The shape of the coil and the method of winding the wire will affect the inductance. Keep in mind that the closer the turns and the greater the alignment of the various layers of wire, the greater the magnetic field that can be produced for a given amount of current and therefore the greater the inductance.

The type of core material used in the coil also affects the inductance. Core material refers to the type of form on which the wire is wound. Many coils are self-supporting and have no form or core. Such coils are referred to as air-core inductors. The inductance of an air-core coil is strictly a function of the factors we have just discussed. When a core is used however the inductance will be affected. The type of core used will determine whether the inductance is greater or less than that of the coil alone with no core.

Most inductors are made with a core that has magnetic properties. Cores made of iron, steel, nickel or some related alloy can support a magnetic field. Such cores concentrate the magnetic lines of force produced by the coil and therefore increase the intensity of the magnetic field. This increases the amount of induced voltage and therefore the inductance.

It is the permeability of the core material that determines its effect on the inductance of the coil. Permeability (μ) is a number that indicates the ability of a material to support the establishment of magnetic lines of force. The greater the ease with which magnetic lines of force can be set up within the core material, the higher the permeability. Air has a permeability of 1 while a magnetic material like iron has a permeability of 7,000.

All of these factors can be related in a formula that can be used to compute the inductance of a single-layer coil inductor as shown in Figure 4-4.

$$L = \frac{.04\mu \, N^2 \, r^2}{l}$$

L is the inductance in microhenrys, N is the number of turns, r is the radius of the coil in centimeters (1 inch = 2.54 centimeters), l is the coil length in centimeters and μ is the permeability. For example, an air-core coil with a radius of 2 cm, a length of 7 cm and 100 turns has an inductance of:

$$L = \frac{.04(1) \, (100)^2 \, (2)^2}{7} = \frac{1600}{7} = 228.57\mu h$$

Air core $\mu = 1$

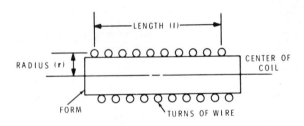

Figure 4-4
Cross-sectional view
of a single layer inductor.

Types of Inductors

There are two basic types of inductors used in electronic circuits, fixed inductors and variable inductors. Fixed inductors are further classified by the type of core material used. The two major kinds of fixed inductors are air-core and iron-core. Variable inductors are inductors whose inductance value can be varied. Most variable inductors use an iron-core.

Fixed Inductors. Basically all inductors are made by winding a length of wire around a core. Solid copper wire with an enamel insulation is the most commonly used conductor. The core can be a non-conducting non-magnetic material such as bakelite, plastic or ceramic. When such cores are used, their primary purpose is to serve as a form or support for the coil of wire. These cores have no magnetic properties and therefore do not affect the inductance of the coil. Where heavy wire is used in making the coil, a form may not actually be needed. The wire may be rigid enough to support itself. Coils with non-magnetic forms or no form at all are referred to as air-core inductors. These coils generally have very low values of inductance and are used primarily in high frequency applications. See Figure 4-5A. Note the schematic symbol used to represent an air-core inductor.

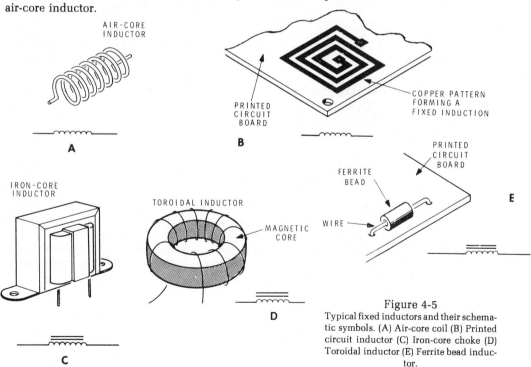

Figure 4-5
Typical fixed inductors and their schematic symbols. (A) Air-core coil (B) Printed circuit inductor (C) Iron-core choke (D) Toroidal inductor (E) Ferrite bead inductor.

Figure 4-5B shows a special type of fixed inductor. The inductor is a spiral of copper foil on a printed circuit board. The advantage of such a coil is that it can be formed when the printed circuit board is being made. No special coil is required. Because of the physical configuration dictated by this type of construction, only very small values of inductance can be obtained. This type of inductor is widely used in high frequency circuits.

Another widely used type of fixed inductor is the iron-core choke. The term choke is used interchangeably with inductor. In this type of inductor, a multi-layer coil is wound on a laminated iron-core. The iron-core which has magnetic properties helps to concentrate the magnetic field produced by the coil thereby increasing its inductance. Very high values of inductance, approaching 100 henrys, are possible with such chokes. Iron-core inductors such as this are used primarily in dc and low frequency ac applications. Refer to Figure 4-5C which shows a typical iron-core choke and its schematic symbol.

Another widely used iron-core choke is the toroidal inductor shown in Figure 4-5D. The donut-shaped core is either a powdered iron called ferrite or a spiral wound tape of magnetic metal. Toroids are used in both low and high frequency applications.

A special iron-core fixed inductor is the ferrite bead inductor shown in Figure 4-5E. Here, a tiny cylindrical-shaped ferrite bead is simply slipped on a wire conductor. Even the short wire is an inductance, but the ferrite bead helps concentrate and intensify the magnetic field around the wire, thereby increasing the inductance. Only small values of inductance can be obtained with a bead, but such small inductors are very useful in high frequency applications.

Variable Inductors. A variable inductor is one whose inductance can be changed. Most variable inductors consist of a coil of wire wound on a non-magnetic form. Inside the form a movable core is placed. The core, usually made of a powdered iron known as ferrite, is made adjustable. The position of the core with respect to the coil can be changed so that the inductance value varies as the core is moved into or out of the coil form. Variable inductors such as this are widely used in tuned circuits in radio applications. Figure 4-6 shows a typical variable inductor and its schematic symbol.

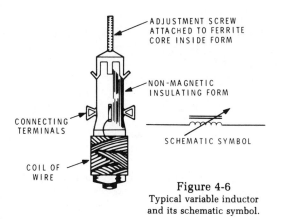

Figure 4-6
Typical variable inductor
and its schematic symbol.

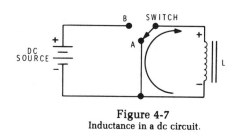

Figure 4-7
Inductance in a dc circuit.

Inductors in DC Circuits

An inductor has absolutely no effect upon direct current unless the dc is changing. In most dc circuits, the current flowing in the circuit is a constant value. When an inductor is used in such a circuit, only the resistance of the wire affects the current. The property of inductance depends upon a changing current which produces a self-induced voltage. When the current in a dc circuit changes, the inductance affects it. The current in a dc circuit usually changes only when it is applied or removed.

Figure 4-7 shows an inductor L connected to a dc source through a switch. When the switch is in position A, the inductor is disabled and no dc is applied to it. When the switch is moved to position B, the dc voltage from the battery is connected to the inductor. When the switch is moved from position A to position B, the dc source voltage will cause current to flow. The instant the electrons begin to flow, a magnetic field will be developed. As the magnetic field expands outward, it cuts the turns of the inductor and induces a counter emf. The polarity of the induced voltage opposes the applied voltage, therefore the amount of current flowing in the circuit is initially limited. Because of the opposition of the induced voltage, the current in the circuit will not rise instantaneously. Instead, it takes a finite period of time for the current to rise to its maximum value. The graph in Figure 4-8A shows how the current varies with time when a dc voltage is first applied to the inductor. Once the magnetic field stops expanding, no further counter emf will be induced. It is at this time that the total current in the circuit will be the maximum value as determined by the amplitude of the applied voltage and the resistance of the inductor.

167

When the switch is now moved from position B to position A, the connection to the dc source is broken. This means that the magnetic field in the inductor collapses. As it collapses, it induces a voltage as the magnetic field cuts the turns of the coil. When the dc source voltage is removed, current does not drop to zero instantaneously. Instead the collapsing magnetic field induces a voltage which causes the current flow to continue in the same direction. When the magnetic field collapses completely, no further voltage will be induced and the current flow will drop to zero. Figure 4-8B shows the decay in current when the power is removed from the inductor.

As you can see from the graphs in Figure 4-8, the inductor clearly opposes changes in current in the dc circuit. When a voltage is applied to the inductor, the inductor itself opposes the rise of current in the circuit. When the voltage is removed from the circuit, the inductor again opposes the change in current flow. This time it tends to keep the current flowing.

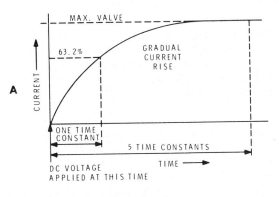

Figure 4-8
Variation of direct
current in an inductor
when (A) power is applied
and (B) power is removed.

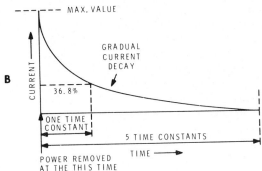

Inductive Time Constant

Whenever a dc voltage is applied to or removed from an inductor, it takes a definite period of time for the current to rise or fall. The effect is somewhat similar to the time that it takes for a capacitor to charge or discharge to the applied voltage. The time that it takes the current to rise or fall is referred to as the time constant of the inductor.

The time constant of an inductor (t) is defined as the time required for the current to rise to 63.2% of its maximum value or to decrease to 36.8% of its maximum value. The time constant is illustrated in Figure 4-8.

The value of the inductive time constant is a function of the inductance and resistance in the circuit. The time constant is directly proportional to the inductance and inversely proportional to the resistance. The simple equation below expresses this relationship.

$$t = \frac{L}{R}$$

As you can see, the greater the inductance, the greater the time constant. The greater the inductance, the greater the induced voltage and therefore the greater the opposition to the applied voltage. This means that the current in the circuit will take longer to rise to its final or maximum value. The larger the resistance, the less current and the lower the strength of the magnetic field and induced voltage.

In this expression, t is the time constant in seconds, L is the inductance in henrys and R is the resistance in ohms. For example, if a 1.5-henry choke has a resistance of 100 ohms, the time constant is:

$$t = \frac{1.5}{100} = .015 \text{ seconds or 15 milliseconds}$$

This means that it takes 15 milliseconds for the current to rise to 63.2% of its maximum value or to fall to a value of 36.8% of the maximum circuit current. It takes approximately 5 time constants for the current to rise to the maximum value from zero or drop from its maximum value to zero.

If the maximum current in the example above is 600 milliamperes, it will take 15 milliseconds for the current to rise to 63.2% of 600 or .632 × 600 = 379.2 mA. It will take 5 × 15 = 75 milliseconds for the current to rise to the full 600 mA.

INDUCTORS IN AC CIRCUITS

When an inductor is used in an ac circuit, it offers opposition to the current flow. The varying applied voltage causes a counter emf to be continuously induced into the coil. The counter emf opposes the applied voltage, and the effect is to limit the amount of current flowing in the circuit. Like a capacitor or resistor, the inductor opposes the flow of alternating current.

Because of the unique relationship between the applied voltage, the induced voltage, and the current in an inductive circuit, an inductor introduces a phase shift. The applied voltage and current will be out of phase with one another because of the inductance. You will study the current-voltage relationships of an inductive ac circuit in this section.

Current-Voltage Relationship

Assume that a sine-wave ac voltage is applied to an inductor as shown in Figure 4-9. Since the applied voltage is sinusoidal, the current flowing in the inductor will also be sinusoidal. The current through the coil will periodically reverse its direction of flow as the applied voltage varies.

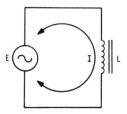

Figure 4-9 Sine-wave ac voltage applied to an inductor.

The current in this inductive circuit is illustrated by the current (I) waveform in Figure 4-10. The changing current in the circuit causes the magnetic field around the inductor to rise and fall, repeatedly cutting the turns of the coil and inducing a voltage. This induced voltage, which we call the counter emf, opposes the applied voltage. The counter emf waveform is also illustrated in Figure 4-10. Note that the counter emf is out of phase with the current sine wave. As you recall from the previous discussion of self-induction, the amount of voltage induced into a conductor is proportional to the rate of change of the magnetic field. The faster the magnetic field expands or collapses, the higher the induced voltage. As you can see from Figure 4-10, the most rapid change in current occurs at points A and B (the zero crossings) where the current is changing from positive to negative or from negative to positive. The rate of change of current is highest at these points. Note that it is at these points where the induced voltage is the highest. The peak positive and negative points of the counter emf occur at the zero crossing points of the current waveform. The rate of change of current is zero at the maximum positive and negative points on the current waveform. It is at these peaks that the current reverses its direction and its rate of change slows to zero. As a result, the induced voltage at these points is also zero. Because of this, the counter emf is 90° out of phase with the current.

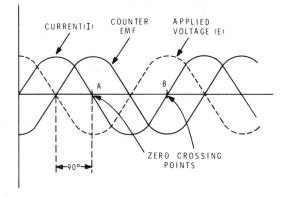

Figure 4-10
Current-voltage relationship
in an inductive circuit.

171

As indicated earlier, the counter emf opposes the applied voltage. Since the induced voltage is in direct opposition, it is exactly opposite in phase to the applied voltage. We say that the applied voltage and the counter emf are 180 ° out of phase with one another. Keep in mind that one cycle of a sine wave represents 360 °. One half cycle represents 180 °, and one quarter of a cycle is 90 °. As you can see from Figure 4-10, the induced and applied voltages are exactly opposite one another. While one is maximum, the other is minimum and vice-versa. Also note that the current lags the applied voltage. As the applied voltage increases, the current increases 90 ° later in time. If the applied voltage decreases, the current decreases 90 ° later. In a purely inductive circuit, the current lags the applied voltage or the applied voltage leads the current by 90 °.

Inductive Reactance

The counter emf induced into an inductor by a varying current opposes the applied voltage. As a result, the total effective voltage in the circuit is the difference between the applied voltage and the induced voltage. Because the induced voltage is less than the applied voltage, the effect of the inductance is to minimize or reduce the current flow. The greater the inductance the greater the counter emf and therefore the higher the opposition to current flow. The opposition to current flow offered by an inductor in an ac circuit is called the inductive reactance. Like resistance and capacitive reactance, inductive reactance is measured in ohms.

The amount of inductive reactance offered by a coil is directly proportional to the inductance and the frequency of the applied voltage. The greater the inductance the greater the magnetic field and the higher the induced voltage for a given current. The higher the induced voltage, the greater the inductive reactance.

As the frequency of the applied ac voltage increases, the rate of change of current also increases. As the rate of change of current increases, the magnitude of the induced voltage becomes greater. This causes greater opposition to the applied voltage.

Inductive reactance is represented by the symbol X_L. The inductive reactance in ohms is expressed by the formula given below:

$$X_L = 2\pi fL$$

$$2\pi = 6.28$$

$$X_L = 6.28\ fL$$

172

In this formula, X_L is in ohms, f is the frequency in Hz and L is the inductance in henrys. For example, to compute the inductive reactance of a .25-henry coil at 400 Hz you would make the following calculations:

$$X_L = 6.28 \ fL$$

$$X_L = 6.28 \ (400) \ (.25)$$

$$X_L = 628 \ ohms$$

By simple algebra, the basic inductive reactance formula can be rearranged to solve for frequency or inductance.

$$f = \frac{X_L}{6.28 \ L}$$

$$L = \frac{X_L}{6.28 \ f}$$

Ohm's Law in Inductive Circuits

Ohm's law applies equally to inductive ac circuits as it does to resistive or capacitive circuits. The current flowing in an inductive ac circuit (I) is directly proportional to the applied voltage (E) and inversely proportional to the inductive reactance (X_L). This relationship is represented mathematically by the expression:

$$I = \frac{E}{X_L}$$

In this expression the current is in amperes, the voltage in volts and the reactance in ohms. Increasing the voltage or decreasing the reactance will cause an increase in the current. Decreasing the applied voltage or increasing the reactance will cause a decrease in the current.

This relationship between the current, voltage and reactance is best illustrated with an example. What is the current flowing in a 2.5-millihenry choke when a 10-volt 100 kHz sine wave is applied to it? To solve this problem, the inductive reactance must first be computed.

$$X_L = 6.28 \ fL \qquad\qquad 100 \ kHz = 100,000 \ Hz$$

$$X_L = 6.28 \ (100,000)(.0025) \qquad 2.5 \ mh = .0025 \ h$$

$$X_L = 1570 \ ohms$$

Now that the inductive reactance and the applied voltage are known, Ohm's law can be used to compute the circuit current.

$$I = \frac{E}{X_L} = \frac{10}{1570} = .00637 \ amperes \ or \ 6.37 \ mA$$

Mutual Inductance

The term inductance is more accurately described as self-inductance. This is the property of a coil whose magnetic field induces a voltage into itself. The varying current in the coil produces a changing magnetic field. The magnetic lines of force cut the turns of the coil producing the magnetic field thereby inducing a voltage into itself that opposes the applied voltage. This phenomenon is known as self-induction.

The varying magnetic field surrounding an inductor can also influence other nearby inductors. As the magnetic field varies, the lines of force can cut the turns of an adjacent coil and induce a voltage into it as well as into the inductor producing the field. The process by which one inductor causes a voltage to be induced into another is called mutual induction.

Figure 4-11 shows an inductor (the primary coil) connected to a source of ac voltage. A varying magnetic field is produced around this coil. When current flows in the primary coil a varying magnetic field is produced. This causes a voltage to be induced into the primary coil itself as well as into the secondary coil. The magnetic lines of force cut the turns of an adjacent inductor (the secondary coil) and induce a voltage into it. The induced voltage causes current to flow through the load resistor. When current begins to flow in the secondary coil a varying magnetic field is produced around it as well. This varying magnetic field induces another voltage into itself. The varying magnetic field from the secondary coil also induces a voltage back into the primary winding. As you can see, there are four induced voltages that affect the operation of the circuit. These are the two self-induced voltages and the two mutually-induced voltages. The overall effect is a complex one because of the interrelationship between the magnetic fields.

The two coils in Figure 4-11 have mutual inductance because one coil induces a voltage into the other. Mutual inductance is designated by the symbol L_m, and the unit of mutual inductance is the henry. A mutual inductance of one henry is defined as the condition where a current change of one ampere per second in one coil induces a voltage of one volt into another coil.

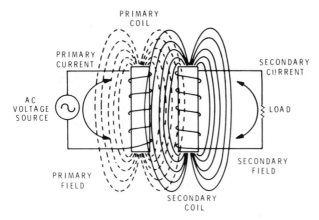

Figure 4-11
Mutual inductance
between adjacent coils.

The amount of mutual inductance that exists between two adjacent coils depends upon the degree of coupling between them. In other words, mutual inductance is determined by how many lines of force of one coil cut the turns of the adjacent coil. The degree of coupling between the coils is expressed by a factor called the coefficient of coupling (k). A coefficient of coupling of one, (k = 1) represents 100% coupling of the magnetic lines of force.

When all of the magnetic lines of force produced by one coil do not cut all of the turns of the other coil, the coefficient of coupling is less than 1. The worst case situation is where none of the magnetic lines of force produced by one coil cut the turns of the other coil. In this case, the coefficient of coupling is zero (k = 0). A coefficient of coupling of zero is obtained when the two coils involved are spaced far enough apart so that their magnetic fields do not influence one another. When two coils are physically mounted so that they are at right angles to one another, the magnetic lines of force of one coil will not influence the other. For other combinations of spacing and relative position of the coils, a coefficient of coupling between 0 and 1 will be obtained.

The mutual inductance that exists between two coils is a function of the coefficient of coupling (k) and the values of inductance of the two coils (L_1 and L_2). The mutual inductance (L_m) is expressed by the equation given below:

$$L_m = k\sqrt{L_1 L_2}$$

The characteristic of mutual inductance only occurs where the magnetic lines of force produced by one coil can affect another adjacent coil. In some situations no mutual inductance is possible. For example, when two iron-core chokes are used the coefficient of coupling between them is zero or very close to it. The iron-core in an inductor concentrates the lines of force and prevents them from extending outward beyond the coil. This is particularly true in iron-core chokes that have a completely closed loop core. None of the magnetic lines of force can leak out of the core and affect adjacent coils. The property of mutual inductance is generally associated with coils mounted on a common core or closely spaced air-core coils.

Inductors in Series and In Parallel

Often inductors are connected in series or parallel combinations to produce different values of inductance. The rules for determining the total inductance of series and parallel connected inductors is similar to the rules for computing the total resistance of resistors in series and in parallel.

Series Inductances. When two or more conductors are connected in series, the total inductance of the combination is simply the sum of the individual inductances. Figure 4-12 shows three inductors connected in series. The total inductance of the combination is:

$$L_T = L_1 + L_2 + L_3$$

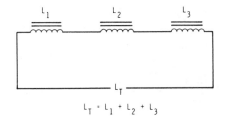

$$L_T = L_1 + L_2 + L_3$$

Figure 4-12
Inductors in series.

The above calculation assumes that no mutual inductance exists between the inductors. If mutual inductance is involved, then the total inductance of the combination will be determined by the individual inductance values and the value of mutual inductance.

When two inductors are connected in series in such a way that their magnetic fields aid one another, the total inductance of the combination is as expressed by the formula below:

$$L_T = L_1 + L_2 + 2 L_m$$

When two coils are connected in series where their magnetic fields oppose one another, the total inductance of the combination is computed with the formula:

$$L_T = L_1 + L_2 - 2 L_m$$

In both of the above situations where mutual inductance is present, the magnetic lines of force of one coil affect those of the other. But the directions of the magnetic fields are important since they can aid or oppose one another. The directions of the magnetic fields can be determined by knowing the directions in which the coils are wound. If the series connected coils are wound so that their turns are all in the same direction, the magnetic fields add. We say that the magnetic fields are series aiding. When the turns of two series connected coils are in opposite directions, the magnetic fields will oppose one another. There will be some cancellation of the magnetic effect between the two coils. If it is desirable to minimize or reduce the mutual inductance to zero, series connected coils should be positioned so that they are right angles to one another.

Parallel Inductances. When two inductors are connected in parallel as shown in Figure 4-13, the total inductance of the combination is as given by the expression below:

$$L_T = \frac{L_1 \, L_2}{L_1 + L_2}$$

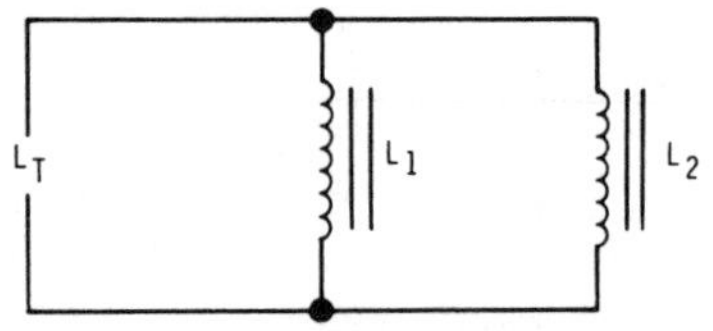

Figure 4-13
Inductors in parallel.

Note that this expression is similar to the formula for parallel resistors or series capacitors. If more than two inductors are connected in parallel, the total inductance of the combination can be determined with the expression above by combining the inductances two at a time until the total is obtained. Again, the expression above assumes that no mutual inductance exists.

Q

Up to this point in our discussion, we have assumed that an inductor is perfect. A perfect inductance or a purely inductive circuit is one that has no resistance. In a purely inductive ac circuit, the only opposition to current flow is the inductive reactance. However, we know from a practical standpoint that a perfect inductor does not exist. Since the inductor is a coil of wire, it has resistance. The resistance of an inductor is simply the resistance of the wire used to make it up. It is not unusual for a choke to have a resistance of several hundred ohms.

In many circuits, the resistance of the inductor is so low that it can be considered negligible. It has little or no effect on the operation of the circuit. However, in other applications where the resistance of the inductor is higher, it can greatly affect the operation of the circuit. Low resistance is a desirable characteristic of an inductor. The primary reason for this is that power is dissipated by a resistance. Therefore, power is dissipated by an inductor because of its resistance. The quality of an inductor therefore is basically determined by its resistance. The quality or figure of merit of an inductor is called the Q. The Q of a coil is the ratio of the energy stored in the coil in the form of a magnetic field to the energy dissipated in the resistance. This can be expressed as indicated below:

$$\text{Quality} = \frac{\text{energy stored}}{\text{energy dissipated}}$$

Since the inductance and inductive reactance is a measure of the amount of energy stored in a coil and since the resistance of a coil is a measure of the amount of energy dissipated, the Q of a coil can also be expressed as the ratio of the inductive reactance to the resistance.

$$Q = \frac{X_L}{R}$$

For example, the Q of a 88 mh coil with a resistance of 50 ohms at 10 kHz is:

$$X_L = 6.28 \, fL \qquad\qquad 88 \text{ mh} = .088 \text{ h}$$

$$X_L = 6.28 \, (10{,}000)(.088) \qquad 10 \text{ kHz} = 10{,}000 \text{ Hz}$$

$$X_L = 5526.4 \text{ ohms}$$

$$Q = \frac{X_L}{R} = \frac{5526.4}{50} = 110.528$$

This relationship is widely used to indicate the quality of an inductor. Generally, the higher the Q the better the coil. A high Q coil is one that has a Q of approximately 20 or greater.

Since Q is directly proportional to the inductive reactance, the Q increases with frequency. The resistance is essentially a constant at low frequencies. But at radio frequencies, the resistance of a coil will increase with frequency. The increase in resistance is a result of skin effect and other high frequency effects. However, the resistance does not increase as rapidly as the inductive reactance. Therefore, the Q still increases with frequency.

RL CIRCUITS

The most commonly used inductive circuit is a series connected resistor and inductor. This combination is called an RL circuit. Even a circuit containing only an inductor is a series RL circuit because of the resistance of the inductor. In this section you are going to analyze the operation and characteristics of a series RL circuit.

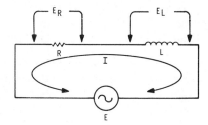

Figure 4-14
A series RL circuit.

Series RL Circuits

Figure 4-14 shows a resistor and inductor connected in series with an ac voltage source. The current flow in the circuit causes voltage drops to be produced across the inductor and the resistor. These voltages are proportional to the current in the circuit and the individual resistance and inductive reactance values. The resistor voltage (E_R) and the inductor voltage (E_L) expressed in terms of Ohm's law are:

$$E_R = IR$$

$$E_L = IX_L$$

For this discussion, we will assume the inductor to be a perfect inductance. That is, the inductor has no resistance or the resistance of the inductor is included with or is represented by the resistor R.

Because this is an inductive circuit, the current flowing in the circuit will lag the applied voltage. In a purely resistive circuit the current and voltage will be in phase. In a purely inductive circuit, the current will lag the applied voltage by 90°. In a series RL circuit, the current will lag the applied voltage by some phase angle between 0° and 90°. The amount of phase shift is a function of the resistance and inductive reactance values.

181

As in any series electrical circuit, the sum of the voltage drops across the various elements in the circuit, equals the applied voltage. In a circuit containing a reactive component such as an inductor or capacitor, the sum of the voltage drops is the vector sum.

Vector Diagram

As in a series RC circuit, a vector diagram can be used to show the relationship between the current and voltages in a series RL circuit. The vectors represent voltage and current amplitudes in the circuit. The direction of the vectors indicates the phase relationships.

Figure 4-15 shows a vector diagram of a simple series RL circuit. The reference vector is labeled I and represents the current in the circuit. The current is common to all circuit elements. The voltage across the resistor is in phase with the current flowing through it. Therefore the resistor voltage vector E_R is shown coinciding with the current vector.

The voltage across the inductor leads the current by 90°. In other words, the circuit current lags the voltage across the inductor by 90°. Therefore, the inductive voltage vector E_L is shown 90° out of phase with the current vector. Recall that we assume that each of the vectors in the diagram is rotating in the counterclockwise direction around the origin (O). The rotation of the vectors represents the variations of voltages and currents in sinusoidal manner over a given period of time. One complete rotation represents one ac cycle. With the counterclockwise direction of rotation, you can see that the inductor voltage is ahead of or leading the current vector. Another way of looking at this is that the current vector is lagging the inductive voltage vector by 90°.

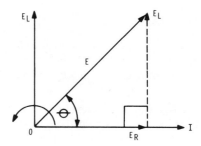

Figure 4-15
Vector diagram of a
series RL circuit.

182

The applied voltage is the vector sum of the resistor and inductor voltages. To find this vector sum, we form a right triangle using the resistor voltage as one side and the inductor voltage as the other side. The applied voltage E is the hypotenuse. Note that the applied voltage leads the current by the angle θ which is between 0° and 90°. Knowing the resistor and inductor voltages, the applied voltage can be found by solving for the hypotenuse of the right triangle formed. This can be done by using Pythagorean's theorem. The formula for the applied voltage is:

$$E = \sqrt{(E_R)^2 + (E_L)^2}$$

By rearranging the above formula, either the resistor or inductor voltages can be computed if the applied voltage and the remaining voltage are known. These formulas are:

$$E_R = \sqrt{(E)^2 - (E_L)^2}$$

$$E_L = \sqrt{(E)^2 - (E_R)^2}$$

Several examples illustrate the use of these formulas.

1. In a series RL circuit the resistor voltage is 15 volts and the inductor voltage is 18 volts. What is the applied voltage?

$$E = \sqrt{(E_R)^2 + (E_L)^2}$$

$$E = \sqrt{(15)^2 + (18)^2}$$

$$E = \sqrt{225 + 324}$$

$$E = \sqrt{549} = 23.43 \text{ volts}$$

2. The ac voltage applied to a series RL circuit is 80 volts. The resistor voltage drop is 32 volts. What is the inductor voltage?

$$E_L = \sqrt{(E)^2 - (E_R)^2}$$

$$E_L = \sqrt{(80)^2 - (32)^2}$$

$$E_L = \sqrt{6400 - 1024}$$

$$E_L = \sqrt{5376} = 73.32 \text{ volts}$$

Impedance

The impedance (Z) of an RL circuit is the total opposition to current flow offered by both the resistor and the inductor. The applied voltage and the total circuit current determine the impedance according to Ohm's law.

$$Z = \frac{E}{I}$$

The impedance is a function of the resistance and the inductive reactance and can be calculated with the formula:

$$Z = \sqrt{(R)^2 + (X_L)^2}$$

The impedance of a series RL circuit is the square root of the sum of the squares of the resistance and the inductive reactance. The impedance, like the resistance and inductive reactance, is expressed in ohms. Again this basic formula can be rearranged to solve for the resistance or the reactance in terms of the other two factors.

$$R = \sqrt{(Z)^2 - (X_L)^2}$$

$$X_L = \sqrt{(Z)^2 - (R)^2}$$

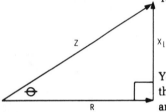

You should recognize that the basic impedance formula is Pythagorean's theorem. Therefore the resistance, inductive reactance and the impedance can be represented by the sides of a right triangle. This is illustrated in Figure 4-16. This vector diagram is similar to the voltage vector diagram given earlier. Since the resistor and inductor voltages are directly proportional to the resistance and inductive reactance, the impedance and voltage vector diagrams are proportional. The angle θ between the impedance (Z) and the resistance (R) is the phase angle of the circuit and indicates the amount by which the circuit current lags the applied voltage. The examples below show the use of the impedance formulas.

Figure 4-16
Impedance triangle
of a series RL circuit.

1. What is the impedance of a 1.5-henry choke with 200 ohms of resistance at 60 Hz?

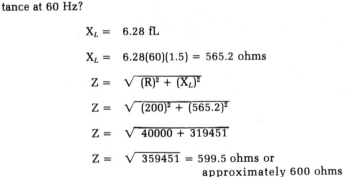

$$X_L = 6.28\ fL$$

$$X_L = 6.28(60)(1.5) = 565.2 \text{ ohms}$$

$$Z = \sqrt{(R)^2 + (X_L)^2}$$

$$Z = \sqrt{(200)^2 + (565.2)^2}$$

$$Z = \sqrt{40000 + 319451}$$

$$Z = \sqrt{359451} = 599.5 \text{ ohms or} \\ \text{approximately 600 ohms}$$

2. In a series RL circuit, the resistor voltage is 20 volts and the applied voltage is 32 volts at 400 Hz. The resistance is 1000 ohms. What is the inductance of the coil?

$$I = \frac{E_R}{R} = \frac{20}{1000} = .02 \text{ amperes}$$

$$Z = \frac{E}{I} = \frac{32}{.02} = 1600 \text{ ohms}$$

$$X_L = \sqrt{(Z)^2 - (R)^2}$$

$$X_L = \sqrt{(1600)^2 - (1000)^2}$$

$$X_L = \sqrt{256000 - 1000000}$$

$$X_L = \sqrt{1560000} = 1249 \text{ ohms}$$

If $X_L = 6.28$ fL then:

$$L = \frac{X_L}{6.28 \text{ f}}$$

$$L = \frac{1249}{6.28(400)} = \frac{1249}{2512} = .496 \text{ h}$$

Phase Shift

The phase shift in a series RL circuit will be some value between 0° and 90°. The actual value of the phase shift is a function of the resistance and inductive reactance. The phase angle θ (theta) is illustrated in Figures 4-15 and 4-16. The phase shift in a series RL circuit is a function of the tangent in the voltage and impedance triangles. As you recall, the tangent is the ratio of the opposite side to the adjacent side. In Figure 4-16 the tangent is the ratio of the inductive reactance (opposite side) to the resistance (adjacent side). Expressed mathematically this is:

$$\tan \theta = \frac{X_L}{R}$$

To solve for the angle itself, we must find the inverse trigonometric function, in this case, the arc tangent. This is expressed mathematically as:

$$\theta = \arctan \frac{X_L}{R}$$

185

Since the current is the same in both the resistor and the inductor, the voltage drops across these elements are directly proportional to the resistance and the reactance. Therefore, the phase angle can also be determined by using the inductor and resistor voltages. This is expressed mathematically as:

$$\theta = \text{arctan} \frac{E_L}{E_R}$$

To find the phase shift in a series RL circuit, determine the resistance and reactance values or the resistor and inductor voltages. Compute the ratio of the inductive reactance to the resistance and look that value up in a table of trigonometric functions. If you use Appendix A in Unit 3, look up the numerical value of the ratio in the Tangent column then locate the closest angle corresponding to that value in the Degrees column.

The example below shows how to compute the phase angle. What is the phase shift produced by a 1.5-henry choke with a resistance of 200 ohms at a frequency of 60 Hz?

$$X_L = 6.28 \text{ fL} = 6.28 \text{ (60) (1.5)} = 565.2 \text{ ohms}$$

$$\theta = \text{arctan} \frac{X_L}{R}$$

$$\theta = \text{arctan} \frac{565.2}{200}$$

$$\theta = \text{arctan} \quad 2.826$$

From Appendix C, tan 70° = 2.748 and tan 71° = 2.904. Therefore the angle corresponding to a tangent of 2.826 is between 70° and 71° or about 70.5°.

Power in an Inductive Circuit

The power dissipated in a series RL circuit is the power dissipated in the resistance. Therefore, to compute the true power dissipation, you can use any of the three standard power formulas:

$$P = EI$$

$$P = I^2R$$

$$P = \frac{E^2}{R}$$

The inductor, like the capacitor, is a reactive component that does not dissipate power in its pure form. The inductor alternately stores energy in the form of a magnetic field and then releases it. During one-half cycle of ac operation, storage of electrical energy occurs in the magnetic field that is built up around the inductor. When that magnetic field collapses, it induces a voltage into the inductor. This self-induced voltage then acts as a source that provides energy to the circuit. The consumption and release of energy cancel one another, thereby making the total average power dissipation zero. Figure 4-17 shows the current, voltage and power relationships in a purely inductive circuit.

Unlike a capacitor, however, an inductor has resistance. For that reason an inductor does dissipate power, but it is the resistive portion of the inductor that causes the power to be dissipated. This is called true power. Figure 4-18 shows the current, voltage and power curves of a series RL circuit where the phase angle is 45° (X_L = R). The power curve above the zero line represents the power dissipated in the resistance (true power) and the power consumed by the inductance. The power curve below the zero line is the power returned to the circuit by the inductance.

To illustrate the power consumption, in a series RL circuit, we will compute the power dissipated in a circuit consisting of a 1000-ohm resistance, an applied voltage of 32 volts at 400 Hz, and an inductance of .5 henry. The total circuit current is .02 amperes.

$$P = I^2R = (.02)^2 \, 1000$$

$$P = .0004 \, (1000) = .4 \text{ watt or } 400 \text{ milliwatts}$$

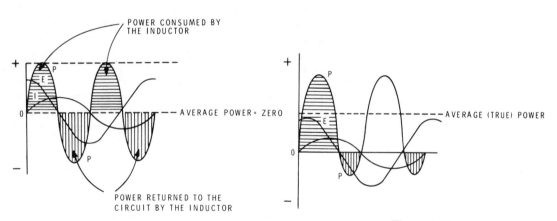

Figure 4-17
Power dissipation
in a pure inductance.

Figure 4-18
Power dissipation in a series RL
circuit with a 45° phase shift.

187

APPLICATIONS OF INDUCTIVE CIRCUITS

Inductive circuits, like capacitive circuits, find wide applications in electronics. The reactive effect of an inductor in an ac circuit makes the inductor valuable in filtering and phase shift applications. However, the inductor is less widely used than the capacitor in such applications. The reason for this is that inductors are larger, heavier, more expensive, come in a narrower range of standard values and dissipate power. This makes the inductor far less valuable as a reactive component than the capacitor in an ac circuit. The greatest advantage of the inductor in ac applications is that it can produce a reactive effect (a change in reactance with a change in frequency) while at the same time completing a dc circuit path. A capacitor can produce a reactive effect but does not offer a complete dc circuit. Often a complete dc path is required as well as a reactive effect.

Inductors are often combined with capacitors to provide improved performance in filtering and phase shift. The reactive effects of capacitors and inductors are opposite one another. As a result, they complement one another in a circuit.

Inductive Filters

Series RL networks can be used as simple low and high-pass filters in the same way that series RC circuits are used. Figure 4-19 shows the two basic types of RL filter circuits. These circuits are resistor-inductor voltage dividers. The circuit in Figure 4-19A is a low-pass filter. The input signal is applied across the coil and resistor in series and the output voltage is taken from across the resistor. At low frequencies the reactance of the coil is low, therefore, very little voltage is dropped across it. Most of the voltage is dropped across the output resistor. As the input frequency increases, the inductive reactance increases. The inductive reactance becomes larger with respect to the resistance and more of the input voltage is dropped across the coil. Therefore less voltage appears across the output resistor. As you can see, increasing the frequency causes an increase in the inductive reactance and a corresponding decrease in the output voltage. Low frequencies are passed with little or no attenuation while high frequencies are greatly reduced in amplitude.

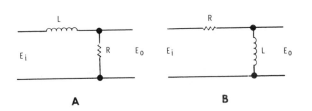

A   **B**

Figure 4-19

Single section RL filters and phase shifters: (A) low-pass filter with lagging phase shift and (B) high-pass filter with leading phase shift.

188

The circuit in Figure 4-19B is a high-pass filter. Again the circuit is a voltage divider with the input ac voltage applied across the coil and resistor in series. The output voltage is taken from across the inductor. At very high frequencies, the inductive reactance is very high compared to the resistance. Most of the input voltage will then appear across the coil and at the output terminals. As the input frequency decreases, the inductive reactance decreases. Less voltage will be dropped across the coil and more across the resistor. The lower the frequency the lower the inductive reactance and therefore the lower the output voltage. High frequencies are passed with little or no attenuation. Low frequencies are greatly attenuated due to the voltage divider effect.

These simple low-pass and high-pass RL filter circuits produce the same result as equivalent RC circuits. But they are less desirable because the inductors are larger and more expensive than capacitors. RC circuits are preferred when such filter networks are required.

The cut-off frequency (f_{co}) of the simple RL networks shown in Figure 4-19 is given by the expression below.

$$f_{co} = \frac{R}{2\pi L} = \frac{R}{6.28 \, L}$$

The cut-off frequency is that frequency above or below which frequencies are passed or attenuated. In the expression above, R is expressed in ohms, L is expressed in henrys and f is expressed in Hz. At the cut-off frequency, X_L equals R and the phase shift is 45°. At the cut-off frequency, the output voltage (E_0) will be approximately 70% of the input voltage (E_i) or $E_0 = .707 \, E_i$. For example, if the input voltage is 8 volts, the output voltage at the cut-off frequency will be $8 \times .707 = 5.656$ volts.

Inductive Phase Shifters

Because an inductor causes the current in a circuit to lag the applied voltage, inductive circuits can be used for phase shifting. The simple series RL circuits shown in Figure 4-19 can also serve as phase shifters. With these circuits, a phase shift between 0° and 90° can be obtained. In the circuit of Figure 4-19A, the output voltage lags the input voltage by some phase angle determined by the inductive reactance and the resistance.

189

The circuit of Figure 4-19B produces a leading phase shift. The output leads the input by a phase angle between 0° and 90°. As the amount of phase shift produced by the circuit approaches 90°, the output voltage becomes greatly attenuated. Theoretically, with a 90° phase shift the output is zero. For that reason such simple RL phase shifters can be used only to provide phase shifts of approximately 60° or less. The greater the phase shift, the greater the need for amplification in the circuit to restore the amplitude of the signal.

The amount of phase shift produced by these circuits is a function of the inductance, the resistance and the input frequency. The phase shift is given by the formula below.

$$\theta = \arctan \frac{X_L}{R} \quad \text{(Figure 4-27A)}$$

$$\theta = \arctan \frac{R}{X_L} \quad \text{(Figure 4-27B)}$$

These formulas assume that the resistance of the inductor is negligible or small compared to the external resistor.

To obtain phase shifts greater than 90°, the simple RL networks in Figures 4-19 can be cascaded. While large values of phase shift can be obtained, the attenuation of the cascaded circuit is considerable. Some type of amplification is generally required to offset the loss in the circuit.

As with RL filters, RL phase shifters are less desirable than RC phase shifters. Inductors are larger, more expensive and have greater losses than capacitors. RC networks are preferred in most cases.

Unit 5

TRANSFORMERS

INTRODUCTION

A transformer is a device which transfers ac electrical energy from one circuit to another. It does this by means of electromagnetic mutual inductance. Generally, the transformer consists of two coils placed close together so that the magnetic field of one coil cuts the other coil. In this way energy is transferred from one coil to the other.

The transformer is the most important single component used in electrical distribution systems. It is also widely used in electronics.

TRANSFORMER ACTION

The transformer is used to transfer alternating current from one circuit to another. Normally, some characteristic of the ac signal is changed in the transformation process. For example, a low voltage ac may be *stepped-up* to a higher voltage ac. Or, a high voltage ac may be *stepped-down* to a lower voltage ac. Often it is the current which must be changed. The transformer can be used to step-up or step-down current. In this section, we will take a look at the action which allows the transformer to do this.

Mutual Inductance

The principle on which transformer action is based is called electromagnetic mutual inductance. This phenomenon was discussed earlier when you studied inductors. We will briefly review the principles of mutual inductance.

Recall that when current flows through a conductor, a magnetic field builds up around the conductor. If alternating current is used, the magnetic field builds, collapses, builds again in the opposite direction, and collapses again for each cycle of the applied current. Another conductor placed in this moving magnetic field will have an emf induced into it.

The transformer is a device which takes advantage of this principle. The two conductors are wound into coils and placed close together so that one coil is cut by the magnetic flux lines of the other.

Figure 5-1 illustrates transformer action. Coil L_1 is connected to an ac voltage source. As alternating current flows through the coil, a varying magnetic field is set up. During one half cycle, current will flow through L_1 in the direction shown. This establishes a north magnetic pole at the top of L_1. As the current increases, the field expands outward cutting the turns of wire in L_2. This induces an emf into L_2 and, in turn, causes current to flow up through the load resistor. Thus, the current in L_1 causes current to flow through L_2.

At the end of the first half cycle, the current through L_1 drops to 0 for an instant when the sine wave input passes through 180°. As the current decreases, the field collapses back into L_1. When the current in L_1 decreases, the current through L_2 also decreases.

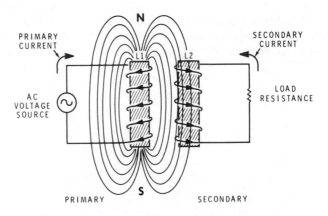

Figure 5-1
Transformer action.

On the next half cycle, the direction of current through L_1 reverses. This causes a magnetic field of the opposite polarity to expand outward from L_1. Once again, this field cuts the turns of L_2 inducing an emf. However, because the polarity of the magnetic field is reversed, the polarity of the voltage induced into L_2 is also reversed. Thus, the induced emf causes current to flow down through the load resistance.

Notice that the current in L_2 follows the current in L_1. Each time the current in L_1 reverses direction, the current in L_2 also reverses. Therefore, the alternating current in L_2 has the same frequency as the alternating current in L_1. Energy is transferred from one circuit to another even though the two circuits are electrically insulated from each other.

The circuit shown in Figure 5-1 is a simple transformer. The coil to which the ac voltage is applied is called the *primary winding*. Current in this winding is caused by the ac voltage source and is called the *primary current*. The coil into which current is induced is called the *secondary winding*. The induced current is called the *secondary current*.

The amount of emf induced into the secondary winding depends on the amount of mutual induction between the two coils. In turn, the amount of mutual induction is determined by the degree of *flux linkage* between the two coils. The flux linkage can be thought of as the percentage of primary flux lines which cut the secondary winding. Another expression which means approximately the same thing is the *coefficient of coupling*. The coefficient of coupling is a number between 0 and 1. When all the primary flux lines cut the secondary coil, the coefficient of coupling is 1. If the two coils are positioned so that some of the primary flux lines do not cut the secondary, then the coefficient of coupling will be less than one.

Figure 5-2 illustrates that the amount of mutual inductance depends on the flux linkage or the coefficient of coupling. In Figure 5-2A, the secondary coil (L_2) is wound directly on the primary coil (L_1). Using this arrangement, nearly all of the flux lines produced by the primary cut the secondary windings. Therefore, the coefficient of coupling is close to one.

In Figure 5-2B the transformer consists of two coils. Here only a few lines of flux from the primary cut the secondary. Therefore, the coefficient of coupling is much lower than in the above example.

Figure 5-2C illustrates that if the two windings are placed far enough apart, there will be no flux linkage between them. In this case, there is no mutual inductance and the coefficient of coupling is zero. While such an arrangement has no practical purpose, it illustrates the importance of the coefficient of coupling.

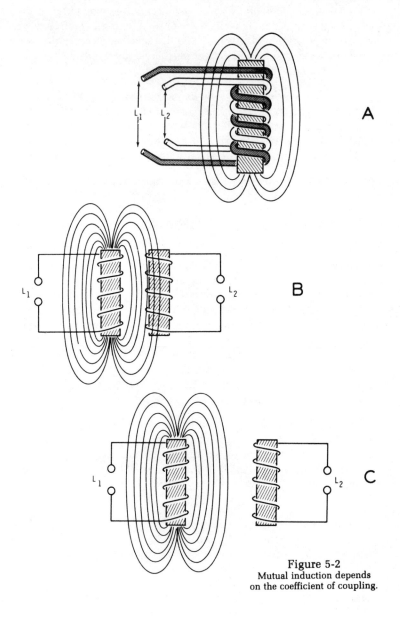

Figure 5-2
Mutual induction depends
on the coefficient of coupling.

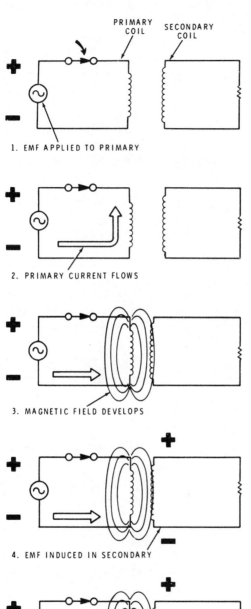

PRIMARY COIL

SECONDARY COIL

1. EMF APPLIED TO PRIMARY

2. PRIMARY CURRENT FLOWS

3. MAGNETIC FIELD DEVELOPS

Figure 5-3 Sequence of events in transformer action.

4. EMF INDUCED IN SECONDARY

5. CURRENT FLOWS IN SECONDARY

Transformer Action

Figure 5-3 illustrates the step-by-step sequence of events which make up transformer action. When the switch is closed the emf of the generator is applied across the primary. This causes current to flow through the primary. The current produces a magnetic field which expands outward, cutting the secondary. This induces an emf in the secondary. If a path for current flow exists, the emf causes a secondary current to flow. The load is connected across the secondary. Energy is transferred from the generator to the load even though the two circuits are not physically connected.

Transformer Construction

The construction techniques used in transformers can vary widely depending on the type of transformer. A substation transformer used in a power distribution system may approach the size of a small house. On the other hand, an IF transformer used in a transistor radio may be no larger than a pencil eraser. In spite of the vast size difference, these two transformers operate on the same basic principle. Both have primary and secondary coils. In both, energy is coupled from the primary to the secondary by mutual inductance.

The design of the transformer is dictated by the frequency that it must pass, the voltage and currents involved and several other factors. A power transformer may be required to handle 115 vac 60 Hz, at 1 ampere. On the other hand, an IF transformer may work with a frequency of 455 kHz at a few millivolts and a few microamperes.

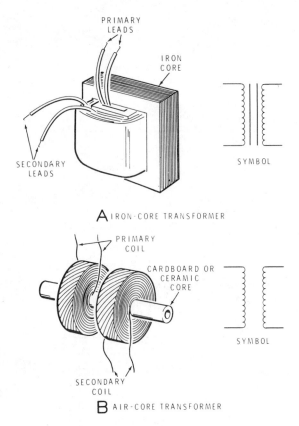

PRIMARY
LEADS

IRON
CORE

SECONDARY
LEADS

SYMBOL

A IRON-CORE TRANSFORMER

PRIMARY
COIL

CARDBOARD OR
CERAMIC
CORE

SYMBOL

SECONDARY
COIL

B AIR-CORE TRANSFORMER

Figure 5-4   Typical
transformer and their symbols.

Figure 5-4 compares the construction of an iron-core transformer with that of an air-core transformer. The iron-core transformer is much larger and heavier. The primary winding is wound on one arm of the core. The secondary is wound directly on top of the primary. Notice that the symbol for the transformer shows the two coils. The two lines between the coils represent the iron core.

The construction of the air-core transformer is different. It is designed to be used at a much higher frequency. Iron core losses increase with frequency. Thus transformers designed to operate at high frequencies use little or no iron in the core. Instead a nonconducting material that has the same permeability as air is used. The core may be ceramic or simply a small cardboard tube.

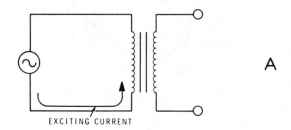

EXCITING CURRENT

A

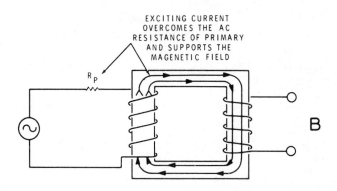

EXCITING CURRENT
OVERCOMES THE AC
RESISTANCE OF PRIMARY
AND SUPPORTS THE
MAGENETIC FIELD

R_P

B

Figure 5-5
Transformer with no load.

TRANSFORMER THEORY

In the previous section you saw how transformer action can be used to couple an ac signal from one circuit to another. In this section, we will explore this action further.

Transformer With No Load

Figure 5-5A shows a transformer being operated without a load. This means that the secondary of the transformer is open. That is, there is no secondary current. Even so, there is some primary current because the primary is connected across an ac voltage source.

The amount of primary current is determined by the impedance of the transformer primary and the applied voltage. Since no power is being taken from the secondary, the primary acts like an inductor. The primary of a typical iron-core transformer can have an inductance of several Henrys. This tends to keep the primary current very low.

In addition to the inductance, the primary winding has a certain value of ac resistance. This tends to limit the current even further. The small amount of primary current that flows with no load is called the exciting current. Figure 5-5B illustrates two functions that the exciting current must perform. First, it overcomes the ac resistance of the primary. In Figure 5-5B, this resistance is shown as a separate resistor. This resistance dissipates power in the form of heat. Secondly, the exciting current supports the magnetic field in the core.

The X_L of the primary is normally much larger than the ac resistance. Thus, the exciting current lags behind the applied voltage by almost 90°. Consequently, when no current flows in the secondary, the primary of the transformer acts like an inductor.

When you studied inductors, you learned that the current lagged the applied voltage because of the counter emf produced by the coil. We will briefly review this principle.

In Figure 5-5B, the applied emf causes current to flow in the primary winding. In turn, this current establishes the magnetic field. However, as the magnetic field expands outward, it cuts the primary winding inducing a counter emf. This counter emf opposes the applied emf. Thus, the current tends to lag behind the applied voltage.

This is the situation which occurs when there is no secondary current. However, when secondary current flows, these conditions are changed and the transformer operates differently. Since the transformer is normally operated with a secondary load, we must understand why it operates differently when secondary current flows.

Transformer With Load

Figure 5-6 shows a simple transformer with a load resistor connected across the secondary winding. When ac flows in the primary, it induces a current into the secondary. Now you will see how the current in the secondary affects the operation of the transformer.

In Figure 5-6A, the polarity of the applied voltage is negative at the top of the primary and positive at the bottom. This forces current to flow down through the primary winding. Using the left-hand rule developed earlier, we find that the current develops a magnetic field in the direction shown.

As this magnetic field expands outward, it induces a counter emf in the primary winding. This counter emf opposes the applied emf. Whereas the applied emf forces current to flow down through the primary, the counter emf tries to force current up through the primary. The net result is a small current which flows down through the primary.

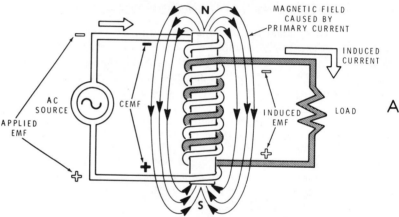

Figure 5-6   Mutual Inductance.

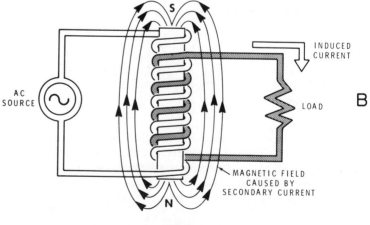

202

Notice that the secondary is wound directly on top of the primary. Therefore, the expanding magnetic field caused by the primary current also cuts the secondary winding. Since the secondary is wound in the same direction as the primary, the emf induced into the secondary has the same polarity as the *counter* emf in the primary. Thus, the induced current in the secondary flows in the direction shown.

The current flow in the secondary establishes a magnetic field of its own as shown in Figure 5-6B. Using the left-hand rule, you can verify that the magnetic field has the polarity shown. As this magnetic field expands, it cuts the secondary winding inducing a counter emf. This counter emf tries to force current to flow down through the secondary in opposition to the induced current.

The expanding flux in the secondary also cuts the primary turns. This induces yet another emf back into the primary winding. This induced emf is in the same direction as the counter emf of the secondary. Thus, this emf tends to force current to flow down through the primary. If you have been keeping track of the various emf's you will see that the emf induced into the primary from the secondary opposes the counter emf originally developed in the primary. Or stated another way, the current induced into the primary (from the secondary) aids the original primary current. This causes the primary current to increase.

This increase in primary current is caused by the expanding magnetic field of the secondary. The more current that flows in the secondary, the stronger the secondary magnetic field will be. This, in turn, increases the primary current. Consequently, an increase in secondary current causes an increase in primary current. Later, you will see that the exact amount of increase in each will depend on the turns ratio.

The sum of the effects described above is called mutual inductance. The inductance is said to be mutual because the primary induces a voltage into the secondary and, simultaneously, the secondary induces a voltage back into the primary.

To be certain you have the idea, review this process once more.

Step 1. AC in the primary establishes a fluctuating magnetic field.
Step 2. The varying flux induces a counter emf into the primary and an emf into the secondary.
Step 3. The induced emf causes current to flow in the secondary.
Step 4. The current in the secondary establishes a magnetic field which is opposite to the field caused by primary current.
Step 5. The secondary flux induces an emf back into the primary which opposes the counter emf of step 2. This decreases the primary counter emf.
Step 6. Primary current increases because the counter emf decreases.

203

TRANSFORMER RATIOS

Transformers have many applications. They are used to step-up or step-down voltage. Also, they are used to step-up or step-down current. Or, they can be used to make one value of impedance appear to be another value. In each case, we are concerned with a ratio. In the first case, the ratio is that of an input voltage to an output voltage. In the second case, the ratio is that of a primary current to a secondary current. In the third case, the ratio is that of an input impedance to an output impedance. As you will see, each of these ratios is determined by the turns ratio of the transformer.

Voltage Ratio

Transformers are frequently used to step-up or step-down voltages. Most electronic devices are powered by 115 VAC, 60 Hz. Some devices require voltages higher than this, while others can get by with much lower voltages. The transformer is used to transform the 115 VAC to whatever value is required.

When the output (secondary) voltage is higher than the input (primary) voltage, the transformer is called a *step-up* transformer. The amount of step-up is determined by the *turns ratio* of the transformer.

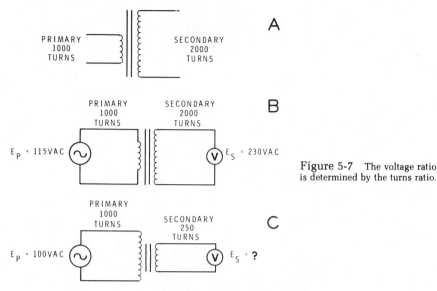

Figure 5-7 The voltage ratio is determined by the turns ratio.

Figure 5-7A illustrates the turns ratio of a typical transformer. Notice that the primary consists of 1000 turns of wire while the secondary has 2000 turns. The turns ratio can be defined as the ratio of the number of turns in the secondary (N_s) to the number of turns in the primary (N_p) that is:

$$\text{Turns ratio} = \frac{N_s}{N_p}$$

Thus in Figure 5-7A, the turns ratio is:

$$\text{Turns ratio} = \frac{N_s}{N_p} = \frac{2000}{1000} = 2$$

Generally this is expressed as a turns ratio of "2 to 1" sometimes written 2:1.

When the secondary has more turns, the voltage is "stepped-up" in proportion to the turns ratio. For example, if the turns ratio is 2:1, then the secondary voltage will be twice as high as the primary voltage. Thus, the voltage ratio is equal to the turns ratio. Expressed as an equation:

$$\frac{E_s}{E_p} = \frac{N_s}{N_p}$$

In some cases, it is more convenient to think of the turns ratio as $\frac{N_p}{N_s}$.

If we do this, the voltage ratio equation is:

$$\frac{E_p}{E_s} = \frac{N_p}{N_s}$$

We can use either of these equations to find the secondary voltage when the turns ratio and the primary voltage are known. Figure 5-7B shows 115 vac applied to the primary. We find the secondary voltage by rearranging the formula:

$$\frac{E_s}{E_p} = \frac{N_s}{N_p}$$

$$E_s = \frac{N_s}{N_p} \times E_p$$

$$E_s = \frac{2000}{1000} \times 115V$$

$$E_s = 2 \times 115V$$

$$E_s = 230V$$

By choosing the proper turns ratio, the input voltage can be stepped-up to any value required.

The transformer can also be used to step-down a voltage. To accomplish this, the secondary should have fewer turns than the primary. For example, in Figure 5-7C, the primary has 1000 turns while the secondary has 250 turns. The primary voltage is given as 100 VAC. You can find the secondary voltage as follows:

$$\frac{E_s}{E_p} = \frac{N_s}{N_p}$$

$$E_s = \frac{N_s}{N_p} \times E_p$$

$$E_s = \frac{250}{1000} \times 100V$$

$$E_s = \frac{1}{4} \times 100V$$

$$E_s = 25V$$

The above equations hold generally true as long as the coefficient of coupling is high and the transformer losses are low. To be completely accurate, the transformer must have a coupling coefficient of 1 and an efficiency of 100%. While these conditions are impossible to achieve in practice, some transformers approach this ideal. In this section, you will assume that an ideal transformer is used. Later, you will see that actual transformers fall short of this ideal.

Power Ratio

If we ignore the losses in the transformer, the power in the secondary is the same as the power in the primary. Thus, in the ideal transformer the power ratio is 1. Although the transformer can step-up voltage, it cannot step-up power. We can never take more power from the secondary than we put in at the primary. Thus, when a transformer steps-up a voltage, it steps-down the current so that the output power is the same as the input power. Expressed as an equation:

$$P_p = P_s$$

where P_p is the power in the primary and P_s is the power in the secondary.

Current Ratio

A transformer which steps-up voltage must at the same time step-down current. Otherwise, it would deliver more power in the secondary than is required at the primary.

We can prove this by deriving an equation for the current ratio. Remember that, ignoring losses:

$$P_p = P_s$$

Recall that the formula for power is: $P = EI$. Thus the power in the primary is equal to $E_p \times I_p$ while the power in the secondary is equal to $E_s \times I_s$. Therefore, if $P_p = P_s$ then:

$$E_p \times I_p = E_s \times I_s$$

The voltage ratio equation was:

$$\frac{E_s}{E_p} = \frac{N_s}{N_p}$$

Transposing:

$$E_s = \frac{N_s}{N_p} \times E_p$$

Substituting this expression for E_s in the previous equation we find:

$$E_p \times I_p = \frac{N_s}{N_p} \times E_p \times I_s$$

Dividing both sides by E_p, we find:

$$\frac{\not{E}_p \times I_p}{\not{E}_p} = \frac{\dfrac{N_s}{N_p} \times \not{E}_p \times I_s}{\not{E}_p}$$

$$I_p = \frac{N_s}{N_p} \times I_s$$

Dividing by I_s, we get:

$$\frac{I_p}{I_s} = \frac{N_s}{N_p}$$

This states that the current ratio is inversely proportional to the turns ratio.

Figure 5-8A shows a transformer with a turns ratio of 4:1. This means that the secondary has 4 times as many turns as the primary. Thus, the voltage is stepped-up by a factor of four, from 10V to 40V. However, the current is stepped-down from 1 ampere in the primary to only 0.25 amperes in the secondary. We can prove this, by rearranging the current ratio formula:

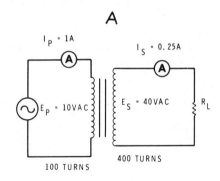

A

$$\frac{I_p}{I_s} = \frac{N_s}{N_p}$$

Cross multiplying:

$$I_s (N_s) = I_p (N_p)$$

Dividing by N_s:

$$I_s = \frac{I_p (N_p)}{N_s}$$

$$I_s = \frac{N_p}{N_s} \times I_p$$

$$I_s = \frac{100}{400} \times 1A$$

$$I_s = \frac{1}{4} \times 1A$$

$$I_s = 0.25A$$

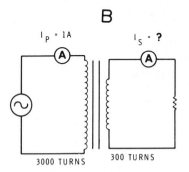

B

Figure 5-8   The turns ratio determines the current ratio.

A transformer can also be used to step-up current. However, in doing this, it must step-down voltage. To step-up current, the primary must have more turns than the secondary as shown in Figure 5-8B. The secondary current is:

$$I_s = \frac{N_p}{N_s} \times I_p$$

$$I_s = \frac{3000}{300} \times 1A$$

$$I_s = 10 \times 1A$$

$$I_s = 10A$$

Solving Transformer Problems

Once you understand how to use the voltage and current ratio formulas, you can solve a wide variety of transformer problems. For example, consider the circuit shown in Figure 5-9A. The number of turns, I_s, and the value of R are given. We wish to find E_p and I_p. We can determine I_p immediately since I_s and the turns ratio are given:

$$\frac{I_p}{I_s} = \frac{N_s}{N_p}$$

$$I_p = \frac{N_s}{N_p} \times I_s$$

$$I_p = \frac{500}{100} \times 10 \text{ mA}$$

$$I_p = 5 \times 10 \text{ mA}$$

$$I_p = 50 \text{ mA}$$

Since we know the turns ratio, we can find E_p if we know E_s. Obviously:

$$E_s = I_s (R_1)$$

$$E_s = 10 \text{ mA} \times 1 \text{ k}\Omega$$

$$E_s = 10V$$

Now we can find E_p:

$$\frac{E_s}{E_p} = \frac{N_s}{N_p}$$

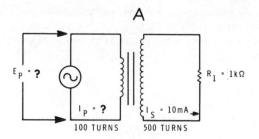

Cross multiplying:

$$E_s (N_p) = E_p (N_s)$$

Dividing by N_s:

$$\frac{E_s (N_p)}{N_s} = E_p$$

Rearranging:

$$E_p = \frac{N_p}{N_s} \times E_s$$

$$E_p = \frac{100}{500} \times 10V$$

$$E_p = \frac{1}{5} \times 10V$$

$$E_p = 2V$$

Figure 5-9
Solving transformer problems.

Figure 5-9B shows another problem. Here, the turns ratio, the primary voltage, and the power dissipated in the secondary circuit are given. We want to find the secondary voltage, the secondary current, the primary current, and the value of R_1.

Because the turns ratio and E_p are given, we can compute the value of E_s:

$$\frac{E_s}{E_p} = \frac{N_s}{N_p}$$

$$E_s = \frac{N_s}{N_p} \times E_p$$

$$E_s = \frac{100}{500} \times 10V$$

$$E_s = \frac{1}{5} \times 10V$$

$$E_s = 2V$$

Once we know E_s, we can compute I_s:

$$P_s = E_s \times I_s$$

$$I_s = \frac{P_s}{E_s}$$

$$I_s = \frac{2W}{2V}$$

$$I_s = 1A$$

When we know I_s and the turns ratio, we can compute I_p:

$$\frac{I_p}{I_s} = \frac{N_s}{N_p}$$

$$I_p = \frac{N_s}{N_p} \times I_s$$

$$I_p = \frac{100}{500} \times 1A$$

$$I_p = \frac{1}{5} \times 1A$$

$$I_p = 0.2A$$

The value of R_1 can now be computed by Ohm's law:

$$R_1 = \frac{E_s}{I_s}$$

$$R_1 = \frac{2V}{1A}$$

$$R_1 = 2\,\Omega$$

or, we find the same value by rearranging the power formula:

$$P = I^2 R$$

$$R = \frac{P}{I^2}$$

$$R = \frac{2W}{1A^2}$$

$$R = 2\,\Omega$$

Impedance Ratio

In electronics, one of the most important applications of a transformer is *impedance matching*. Maximum power is transferred from a generator to a load when the impedance of the generator matches that of the load. If the impedances do not match, a great deal of power can be wasted.

There are many cases in electronics in which the impedance of the signal source (generator) simply does not match that of the load which it must drive. For example, a transistor amplifier stage might be most efficient when driving a 100-ohm load. Nevertheless, the amplifier may be required to drive a 4-ohm speaker. This is the kind of mismatch that results in wasted power and inefficient operation.

Fortunately, a transformer can be used to solve this impedance-matching problem. The transformer can make one value of impedance appear to be another value. In the above example, a transformer can be placed between the transistor amplifier and the speaker. By choosing the proper turns ratio, the transformer can make the 4-ohm speaker appear as a 100-ohm load to the transistor amplifier.

We have seen that the voltage or current step-up of a transformer depends on the turns ratio. The impedance matching capability of a transformer also depends on turns ratio. However, the impedance ratio is equal to the turns ratio *squared*. That is:

$$\frac{Z_p}{Z_s} = \left(\frac{N_p}{N_s}\right)^2$$

Where Z_p is the impedance of the primary circuit; Z_s is the impedance of the secondary circuit; and $\dfrac{N_p}{N_s}$ is the primary to secondary turns ratio. The formula can be rearranged:

$$\frac{N_p}{N_s} = \sqrt{\frac{Z_p}{Z_s}}$$

We can use the equation in this form to solve the impedance-matching problem discussed earlier. The problem is to find a turns ratio which will match a 100-ohm generator (transistor amplifier) to a 4-ohm load (speaker). Using the formula:

$$\frac{N_p}{N_s} = \sqrt{\frac{Z_p}{Z_s}}$$

$$\frac{N_p}{N_s} = \sqrt{\frac{100\Omega}{4\Omega}}$$

$$\frac{N_p}{N_s} = \sqrt{25}$$

$$\frac{N_p}{N_s} = 5$$

This is a primary to secondary turns ratio of 5:1. Thus, if the number of primary turns is 5000, then the number of secondary turns must be 1000. As you can see, a transformer with a turns ratio of 5:1 has an impedance ratio of 5^2:1 or 25:1.

Let's consider another example. Suppose we wish to match a generator which has an impedance of 6000 ohms to a 60-ohm load. Here the proper turns ratio is:

$$\frac{N_p}{N_s} = \sqrt{\frac{Z_p}{Z_s}}$$

$$\frac{N_p}{N_s} = \sqrt{\frac{6000}{60}}$$

$$\frac{N_p}{N_s} = \sqrt{100}$$

$$\frac{N_p}{N_s} = 10$$

That is, the turns ratio must be 10:1, or the primary must have 10 times as many turns as the secondary.

Impedance matching is one of the most important applications of a transformer. By choosing the proper turns ratio, transformers can match a wide range of impedances.

TRANSFORMER LOSSES

Transformers are very efficient devices. An efficiency of ninety percent or better is normal. Nevertheless, the transformer does have some losses. In many cases, these losses dictate the design of the transformer. Power transformers in particular are designed so that losses are minimized. The losses can be broken down into several categories.

Core Losses

In power transformers, the largest loss occurs in the core. Even so, a much larger core loss would occur if it were not for special construction techniques. Core losses can be divided into two separate parts. We will examine each of these losses.

Eddy Current Losses. The cores of power transformers are generally made of soft iron or steel. Because iron and steel are good conductors, a current can be induced into the core when the core is subjected to a moving magnetic field. As you have seen, a moving magnetic field is a requirement in all transformers. Thus, unless special precautions are taken, large circulating currents will be induced into the core of the transformer. These currents are called eddy currents.

Figure 5-10 shows how eddy currents are induced into the core. When alternating current flows through the winding, a changing magnetic field is established in the core. As this field expands and contracts, it induces a voltage into the core. The induced emf causes eddy currents to flow as shown.

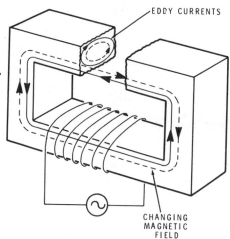

EDDY CURRENTS

CHANGING
MAGNETIC
FIELD

Figure 5-10   Eddy currents are induced in the core by the changing magnetic field.

Eddy currents can be reduced by changing the construction of the core. In Figure 5-10 a solid block of metal is used as the core. Because the cross sectional area of the core is large, it has very little resistance and large eddy currents can flow.

The eddy currents produce a power loss which is proportional to the current squared ($P = I^2 R$). If the eddy currents could be reduced, the power loss would be reduced also.

Eddy currents can be reduced by using many thin sheets of metal for the core rather than using a solid block of metal. Figure 5-11 illustrates how this reduces the eddy currents. Figure 5-11A shows large eddy currents flowing through the low-resistance solid core. However, a core having the same magnetic characteristics can be formed of several thin sheets as shown in Figure 5-11B. The sheets are coated with an insulating varnish so that no current can flow from one sheet to the other. Thus, any eddy currents produced are restricted to a single sheet of metal. Because the cross sectional area of a sheet is quite small, the resistance of each individual sheet is relatively high. This keeps the magnitude of the eddy currents low. Consequently, the power loss is much lower in a core made of thin sheets.

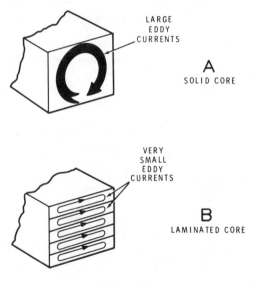

LARGE
EDDY
CURRENTS

A
SOLID CORE

VERY
SMALL
EDDY
CURRENTS

B
LAMINATED CORE

Figure 5-11
Eddy currents can be
reduced by laminating the core.

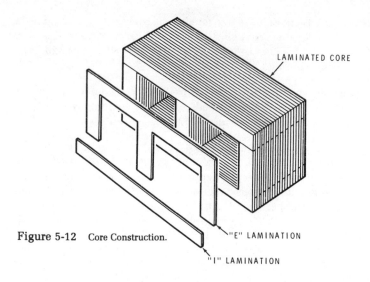

LAMINATED CORE

"E" LAMINATION

"I" LAMINATION

Figure 5-12   Core Construction.

The thin sheets which make up the core are called *laminations*. A laminated iron core is shown in Figure 5-12. The lamination is named according to its shape. Frequently, an iron core is constructed by interleaving "E" and "I" laminations as shown. Remember, the purpose of laminating a core is to reduce the power loss caused by eddy currents.

Hysteresis. Another type of loss which occurs in the core of the transformer is called *hysteresis*. When iron is not magnetized, its magnetic domains are arranged in a random pattern. However, if a magnetizing force is applied, the domains line up with the magnetic field. When the magnetic field reverses, the domains must reverse their direction also. In a transformer, the magnetic field will reverse direction many times each second in step with the applied ac signal. Thus, the domains must reverse their direction many times each second. When reversing direction, the domains must overcome friction and inertia. In doing this a certain amount of power is dissipated in the form of heat. This power loss is referred to as a *hysteresis loss*.

In some materials, the hysteresis loss is quite high. Soft iron normally has a high hysteresis loss. The hysteresis loss in steel is generally lower. Some large power transformers use a special type of metal called silicon steel because it has a low hysteresis loss.

The amount of loss caused by hysteresis increases with frequency. An iron-core transformer which has little loss at low frequencies, may have a large hysteresis loss at higher frequencies.

Copper Loss

Another type of loss present in all transformers is called *copper loss*. This loss is caused by the ac resistance of the copper wire in the primary and secondary windings. A transformer winding can consist of hundreds of turns of fine copper wire. Because of the length of the wire and its tiny cross sectional area, the ac resistance can be quite high. As current flows through this resistance, some power is dissipated in the form of heat. The amount of power can be determined by the formula: $P = I^2 R$. For this reason, another name for copper loss is $I^2 R$ *loss*.

The amount of copper loss is proportional to the current squared. Thus, if the current in the transformer doubles, the copper loss increases by a factor of 4.

The copper loss can be reduced by increasing the size of the copper wire in the windings. Larger wire has less resistance and thus the $I^2 R$ loss is reduced. Another approach is to keep the current in the transformer as low as possible.

External Induction Loss

As the magnetic field expands and contracts around the transformer it often cuts an external conductor of some kind. If a current is induced into the conductor, some power is lost from the transformer circuit. In most cases, the power lost by external induction is so small that it can be ignored. However, the voltage induced into outside circuits can be bothersome. For example, in a sensitive amplifier circuit, the unwanted induced voltage from a transformer may interfere with the signal we are trying to amplify.

Interference caused by induction from a transformer can be reduced by shielding. Often, sensitive circuits are placed inside a metal shield which will prevent stray magnetic fields from reaching the circuits. Also, transformers themselves are often placed in thin metal housings to prevent magnetic fields from escaping.

Transformer Efficiency

Because of the losses mentioned previously, more power is applied to the primary of the transformer than is available for use in the secondary. That is, every transformer has a certain power loss. Thus, the *efficiency* of a transformer is always less than 100%.

The efficiency of a transformer is the ratio of output power to input power. For example, if a transformer has an input power of 110 watts and an output power of 105 watts,

$$\text{efficiency} = \frac{\text{power output}}{\text{power input}} = \frac{105\text{W}}{110\text{W}} = 0.9545$$

However, efficiency is normally given in percent. Thus, the decimal fraction must be multiplied by 100 to convert to percent efficiency. Also, the input power is the primary power while the output power is the secondary power. Thus, the formula for percent efficiency is:

$$\% \text{ efficiency} = \frac{P_s}{P_p} \times 100$$

P_p is the power in the primary or input power. P_s is the power in the secondary or output power. Thus, in the above example:

$$\% \text{ efficiency} = \frac{P_s}{P_p} \times 100$$

$$= \frac{105\text{W}}{110\text{W}} \times 100$$

$$= .9545 \times 100$$

$$= 95.45\%$$

TRANSFORMER APPLICATIONS

Transformers are very versatile devices. They are used to step-up voltage, to step-down voltage, to step-up current, to step-down current and to match impedance. Also, they can produce a 180° phase shift, provide two signals which are 180° out of phase, and isolate one circuit from another. Finally they can pass ac while blocking dc and they can provide several different signals at various voltage levels. Next, we will discuss some of these applications in more detail.

Power Distribution

One of the most important applications of transformers is in the transmission of power over long distances. Often the power generating stations are located near coal fields or at dams far from the cities where the electrical power is needed. The electrical energy must be transported over great distance by transmission lines. The transformer plays a major role in transmitting this energy efficiently.

Long distance transmission lines are generally made of aluminum. Copper lines have better electrical characteristics but are too expensive, too heavy, and lack structural strength. You will recall that aluminum has a higher resistivity than copper. Consequently, the power losses in aluminum lines can be high unless special precautions are taken.

Most of the power lost in a transmission line is caused by the resistance of the line. Recall that: $P = I^2 R$. Thus, the amount of power loss is proportional to the resistance of the line and to the *square* of the current. Obviously then, the easiest way to reduce the power losses in the line is to keep the current as low as possible.

For example, suppose a generating station produces 12,000 volts at 10 amperes. Assume that this 120,000 watts of power is transmitted over a transmission line having a resistance of 100 ohms. If transmitted as 12,000V at 10A, the power loss in the line is:

$$P = I^2 R$$

$$P = (10A)^2 \times 100\Omega$$

$$P = 100 \times 100$$

$$P = 10,000 \text{ watts}$$

By using a transformer, the 120,000 watts can be transmitted as 120,000 V at 1A. In this case, the power loss is:

$$P = I^2 R$$

$$P = (1A)^2 \times 100\Omega$$

$$P = 1 \times 100$$

$$P = 100 \text{ watts}$$

Notice that by stepping the current down by a factor of 10, the power loss is reduced by a factor of 100. For this reason, power is transmitted over great distances at very high voltage levels and very low current levels. Upon reaching its destination, the electrical power is stepped-down in voltage to the values required by homes and industry.

Electronic Applications

In electronic devices the transformer is also used to step-up or step-down voltage. Many electronic devices require 115 vac for power. Most of the devices have a power transformer which steps the voltage up or down as required. In transistorized equipment, the ac line voltage is generally stepped-down and then changed to dc. In older vacuum-tube equipment, the line voltage was first stepped-up and then changed to dc. Vacuum tubes require a higher dc voltage than transistors. As you can see, the transformer can be used to make the line voltage compatible with both types of equipment.

Transformers are also used as impedance-matching devices. They can match the impedance of one circuit to that of another. The impedance must be matched if maximum power is to be transferred between the two circuits.

The above applications were discussed in detail earlier. Now, take a look at some additional applications.

Phase Shifting. Depending on how the transformer is wound, it will provide either a 180° phase shift or no phase shift. That is, the voltage in the secondary will be either in phase or 180° out of phase with the voltage in the primary.

In some applications the phase shift is unimportant while in other applications it is extremely important. Figure 5-13A shows a transformer in which the input signal is in phase with the output signal.

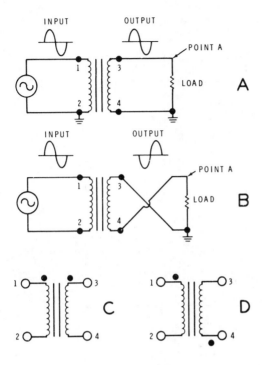

Figure 5-13
Phase relationships in transformers.

The voltage at point A with respect to ground has the same phase as the voltage at pin 1 with respect to ground. If we wish to have a 180° phase shift between input and output, we must reverse the leads to the load. Figure 5-13B shows the leads reversed. Notice that this places ground at pin 3 of the transformer. The voltage at point A is now 180° out of phase with the input voltage.

The phase relationship between the windings of the transformer are sometimes indicated on schematic diagrams by dots (•) as shown in Figure 5-13C. The ends of the windings which are marked by dots are at the same phase. Thus, the voltage at pin 3 is in phase with the voltage at pin 1. Figure 5-13D shows a transformer that is wound differently. Here the voltage at pin 4 is in phase with the voltage at pin 1. Or stated another way, the voltage at pin 3 is 180° out of phase with the voltage at pin 1.

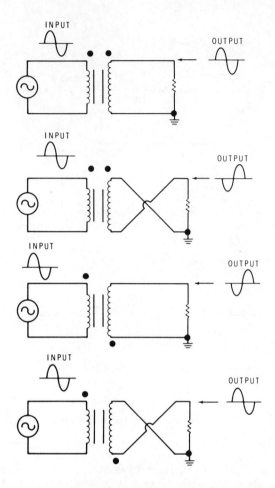

Figure 5-14
The phase can be reversed by
reversing the leads to the load.

Figure 5-14 shows that an in-phase output or 180° out-of-phase output can be obtained from either type of transformer. Notice that the phase can be reversed simply by reversing the leads to the load.

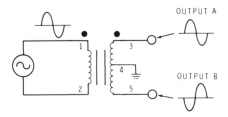

OUTPUT A

OUTPUT B

Figure 5-15   A center-tapped
secondary produces two
signals which are out of phase.

Phase Splitting. Some circuits require two ac signals of equal amplitude which are 180° out of phase. A transformer can be used to provide the two out-of-phase signals. Figure 5-15 shows a transformer with a center-tapped secondary. This simply means that the center of the secondary is brought out to a terminal (pin 4). The dots indicate that the voltage at pin 3 is in phase with the voltage at pin 1. Ignoring the center tap, the voltage at pin 5 must be 180° out of phase with the voltage at pins 1 and 3. If the center tap is grounded, two signals which are equal in amplitude and 180° out of phase exist at the opposite ends of the secondary. Thus, the transformer can be used to produce two signals which are 180° out of phase.

Isolation. Another purpose of a transformer is to isolate one circuit from another. A line operated device which does not use a power transformer often has a metal chassis which connects to one side of the ac line. Anyone touching this chassis and ground at the same time can receive an electrical shock. However, if a power transformer is used, the chassis is isolated from the ac line. Thus, the possibility of accidental shock is greatly reduced.

Frequently, technicians must repair transformerless equipment. When the "hot" chassis is removed from its plastic or wooden cabinet, the possibility of accidental shock increases. To safeguard himself, the technician places an isolation transformer between the ac line and the chassis. The transformer has a turns ratio of 1 to 1. That is, it takes 115 VAC from the line and delivers 115 VAC to the chassis. However, it isolates the chassis from the ac line and protects the technician from accidental shock.

Autotransformer

The autotransformer is a special type of transformer in which there is no isolation between the primary and the secondary. A single continuous coil is wound on a core. Part of this coil is used as the primary while another part is used for the secondary. Generally, part of the coil will be used in both the primary and the secondary.

Figure 5-16A shows the autotransformer being used to step down the applied voltage. Here the entire winding serves as the primary. The lower half of the coil is used as the secondary winding. Because there are fewer turns in the secondary than in the primary, the voltage is stepped down and the current is stepped up.

Figure 5-16B shows that the transformer can be turned around and used to step up voltage. Here the lower half of the coil is used as the primary while the entire coil is used as the secondary. Since the secondary has more turns, the transformer steps up voltage and steps down current.

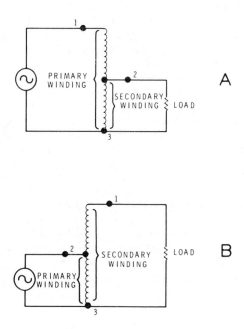

Figure 5-16   The autotransformer.

Figure 5-17 compares the autotransformer with the conventional transformer. The conventional transformer in Figure 5-17A steps the applied voltage down from 120 VAC to 20 VAC. This requires a turns ratio of 6:1. Ignoring losses, the current is stepped up from 1A to 6A.

Figure 5-17B shows how the same job can be accomplished with an autotransformer. The tap at pin 2 is placed at 1000 turns. Thus, the primary (the entire coil) has 6000 turns but only 1000 of these are used as the secondary. Notice that 6 amperes flow in the load but only 5 amperes flow in the secondary winding. The reason for this is that the current in the secondary winding flows opposite to the primary current. Thus, the 1-ampere primary current subtracts from the 6-ampere secondary current.

This illustrates the advantages of the autotransformer. First notice that fewer turns of wire are required in the autotransformer. Also, since the current in the secondary winding is lower, the $I^2 R$ loss is lower. In many cases, the autotransformer is also easier to construct and therefore cheaper. Its prime disadvantage is that the secondary is not isolated from the primary.

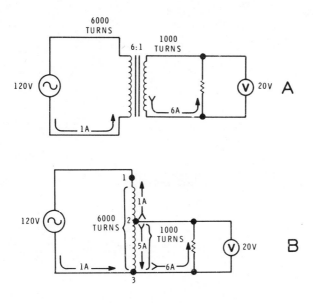

Figure 5-17
Comparing the autotransformer
to the conventional transformer.

225

A special type of autotransformer is shown in Figure 5-18. The load is connected between the movable arm and the bottom of the coil. By moving the arm up or down the turns ratio can be changed. This causes a corresponding change in the voltage across the load. The output voltage can vary from about 0 VAC to over 130 VAC. This device is called a variable transformer.

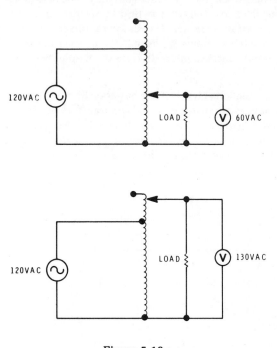

Figure 5-18
The variable transformer.

SUMMARY

The following is a point by point summary of this unit.

The transformer is a device for coupling an ac signal from one circuit to another.

During the transformation process, the voltage or the current may be changed. However, the frequency is never changed.

A transformer consists of two or more windings wound on a core. The winding to which an input signal is applied is called the primary. The winding from which an output signal is taken is called the secondary.

Transformer action is based on electromagnetic mutual inductance. Current in the primary produces a magnetic field which induces an emf into the secondary. In turn, secondary current produces a magnetic field which induces an emf into the primary. This emf opposes the counter emf of the primary. Consequently, an increase in secondary current causes an increase in primary current.

The voltage ratio, the current ratio, and the impedance ratio of a transformer is determined by the turns ratio. The formula for the turns ratio is:

$$\text{Turns ratio} = \frac{N_s}{N_p}$$

The formula for the voltage ratio is:

$$\frac{E_s}{E_p} = \frac{N_s}{N_p} \quad \text{or} \quad \frac{E_p}{E_s} = \frac{N_p}{N_s}$$

The formula for the current ratio is:

$$\frac{I_s}{I_p} = \frac{N_p}{N_s} \quad \text{or} \quad \frac{I_p}{I_s} = \frac{N_s}{N_p}$$

These formulas hold true for a transformer which is 100% efficient. Since all transformers have some losses, the formulas are not always accurate.

A transformer can be used to match the impedance of one circuit to that of another. In many cases this is important since maximum power is transferred only when the impedance of the source matches that of the load. The impedance matching formula is:

$$\frac{Z_p}{Z_s} = \left(\frac{N_p}{N_s}\right)^2 \quad \text{or} \quad \frac{N_p}{N_s} = \sqrt{\frac{Z_p}{Z_s}}$$

No transformer is 100% efficient. All have power losses. The losses consist primarily of core losses and copper losses. Core losses are caused by eddy currents and hysteresis. Many transformer cores are laminated to reduce eddy current losses. Copper losses are caused by the ac resistance of the transformer windings. The efficiency of a transformer can be determined by the formula:

$$\% \text{ efficiency} = \frac{P_s}{P_p} \times 100$$

Some applications of transformers include:
    Stepping up voltage
    Stepping down voltage
    Stepping up current
    Stepping down current
    Matching impedance
    Providing a 180° phase shift
    Providing two signals which are 180° out of phase
    Isolating one circuit from another
    Passing ac but blocking dc

An autotransformer consists of a continuous winding on a core. It may be used like any other transformer except that it does not provide isolation between the input and output.

An isolation transformer is used to isolate one circuit from another. Frequently, a 1:1 isolation transformer is used by technicians to isolate a transformerless chassis from the ac line.

Unit 6

TUNED CIRCUITS

INTRODUCTION

Tuned circuits are used extensively in electronics. They are used to determine the frequency at which oscillators and amplifiers operate, to separate wanted signals from unwanted signals, and to eliminate interference and noise. Without tuned circuits many phases of electronics could not exist.

This unit will serve as an introduction to tuned circuits. It deals with resonance and filters. As you will see, these are extremely important phases of electronics.

RLC CIRCUITS

RLC circuits are circuits which have resistors (R), inductors (L), and capacitors (C) connected together in some manner. The simplest RLC circuit consists of a resistor, an inductor and a capacitor connected in series. This is called a series RLC circuit.

Series RLC Circuits

Figure 6-1 shows a resistor, coil, and capacitor connected in series with an ac generator. Because the values of capacitance, inductance, and frequency are given, the reactances in the circuit can be computed using the formulas:

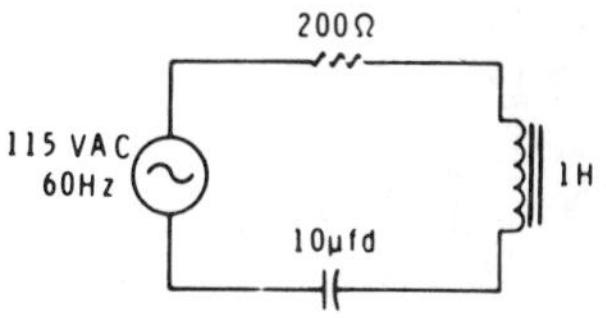

Figure 6-1
Series RLC Circuit

$$X_L = 2\pi\ fL \text{ and } X_C = \frac{1}{2\pi\ fC}$$

The inductive reactance is:

$$X_L = 2\pi\ fL$$
$$X_L = 6.28 \times 60 \text{ Hz} \times 1 \text{ H}$$
$$X_L = 377\ \Omega$$

The capacitive reactance is:

$$X_C = \frac{1}{2\pi\ fC}$$

$$X_C = \frac{1}{6.28 \times 60 \text{ Hz} \times 0.00001 \text{ fd}}$$

$$X_C = \frac{1}{0.00377}$$

$$X_C = 265\ \Omega$$

The reactance values combine with the resistance value to form an impedance to current flow. Recall that there is a general formula for impedance: $Z = \sqrt{R^2 + X^2}$. When the reactance is caused by capacitance, the formula becomes: $Z = \sqrt{R^2 + X_C^2}$. When inductive reactance is involved, the formula becomes $Z = \sqrt{R^2 + X_L^2}$.

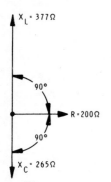

Figure 6-2
Vector Diagram for
Series RLC Circuit

The circuit shown in Figure 6-1 contains both capacitive and inductive reactance. Thus, a formula is needed which contains both X_C and X_L. A vector diagram of the reactances and resistance in the circuit will help develop such an equation. Figure 6-2 shows the vector diagram for this circuit. Following the conventions established earlier, R is plotted at 0; X_L is plotted at $+90°$, and X_C is plotted at $-90°$. Notice that this places X_L 180° out of phase with X_C. For this reason, X_L and X_C tend to cancel. However, because X_L is greater than X_C, the resultant reactance will be inductive in nature.

Figure 6-3 shows the resultant vector. Because X_L was 377 Ω and X_C was 265 Ω, X_L can completely cancel X_C and still have a value of 112 Ω. That is, 377 Ω − 265 Ω = 112 Ω. Thus, when a circuit has both capacitive reactance and inductive reactance, the net reactance is the difference between the two. Because of this, the complete formula for the impedance in a series ac circuit is:

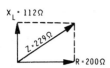

Figure 6-3
Resultant Vector after X_L cancels X_C

$$Z = \sqrt{R^2 + (X_L - X_C)^2}$$

$$Z = \sqrt{(200)^2 + (377 - 265)^2}$$

$$Z = \sqrt{(200)^2 + (112)^2}$$

$$Z = \sqrt{40,000 + 12,544}$$

$$Z = \sqrt{52,544}$$

$$Z = 229 \ \Omega$$

The net result is the same as if a coil having an X_L of 112 Ω is connected in series with a resistance of 200 Ω. Figure 6-3 shows the final vector diagram.

The above formula for impedance is used when X_L is greater than X_C. Obviously, there will be cases in which X_C is larger than X_L. Such a case is shown in Figure 6-4A. Here X_L is 10 ohms; X_C is 40 ohms and R is 40 ohms.

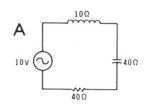

Figure 6-4B shows the vector diagram for this circuit. Notice that in this case X_C more than cancels X_L. Thus, the net reactance is found by subtracting X_L from X_C. As shown in Figure 6-4C, the result is a capacitive reactance of 30 Ω. Consequently, when X_C is larger than X_L, the formula for impedance becomes:

$$Z = \sqrt{R^2 + (X_C - X_L)^2}$$

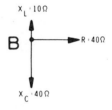

Using this formula, we find that the total impedance of this circuit is:

$$Z = \sqrt{R^2 + (X_C - X_L)^2}$$

$$Z = \sqrt{(40)^2 + (40-10)^2}$$

$$Z = \sqrt{(40)^2 + (30)^2}$$

$$Z = \sqrt{1600 + 900}$$

$$Z = \sqrt{2500}$$

$$Z = 50 \ \Omega$$

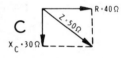

Once the impedance of the circuit is known we can determine other circuit values. Using formulas developed in earlier units, we can compute values such as current, the voltage dropped by each component and the power factor. For example, the current in Figure 6-4A must be:

$$I = \frac{E}{Z} = \frac{10 \ V}{50 \ \Omega} = 0.2 \ \text{amps}$$

Figure 6-4
Circuit and Vectors in which
X_C is greater than X_L

233

This allows us to find the voltage drop across each component.

$$E_R = I \, (R) = 0.2A \times 40 \, \Omega = 8 \; V$$

$$E_L = I \, (X_L) = 0.2A \times 10 \, \Omega = 2 \; V$$

$$E_C = I \, (X_C) = 0.2A \times 40 \, \Omega = 8 \; V$$

The power factor can be determined by the formula:

$$PF = \frac{R}{Z} = \frac{40 \, \Omega}{50 \, \Omega} = 0.80$$

And since the power factor is also equal to the cosine of the angle, we can find the angle from a cosine chart. Using such a chart we find that the angle is about 36.5°. The vector shown in Figure 6-4C shows that this is a negative angle. This means that the circuit acts capacitively.

We can compute apparent power in volt-amps by multiplying the applied voltage times the current: VA = E × I = 10 V × 0.2A = 2 VA. However, the true power is somewhat less since only the resistor can dissipate power:

$$P = I^2R = (0.2A)^2 \times 40 \, \Omega = 0.04 \times 40 = 1.6 \; W$$

The above procedure is a step-by-step way of analyzing a series RLC circuit. Using this procedure, we can compute all important circuit values from a few given values.

Parallel RLC Circuits

Figure 6-5 shows a parallel RLC circuit. Because the three components are in parallel, the same voltage is applied across each. As you learned earlier, it is generally easier to work with currents in problems of this type. Recall that when a reactance is in parallel with a resistance, the total current is the vector sum of the two branch currents. That is,

$$I_T = \sqrt{I_R^2 + I_X^2}$$

In this formula, I_X may be the current through either a capacitor or an inductor.

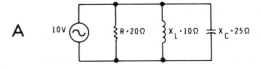

This formula can be expanded to handle situations like that shown in Figure 6-5A where both inductors and capacitors are used. The first step is to find the current through each branch. This is easy to do since all the pertinent values are given. The branch currents are:

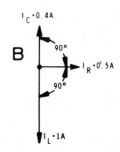

$$I_R = \frac{E}{R} = \frac{10\ V}{20\ \Omega} = 0.5\ \text{amperes}$$

$$I_L = \frac{E}{X_L} = \frac{10\ V}{10\ \Omega} = 1\ \text{ampere}$$

$$I_C = \frac{E}{X_C} = \frac{10\ V}{25\ \Omega} = 0.4\ \text{amperes}$$

Figure 6-5B shows the branch currents plotted as vectors. I_R is in phase with the applied voltage and is plotted at 0°. I_L lags the applied voltage by 90° while I_C leads the applied voltage by 90°. Notice that I_C is 180° out of phase with I_L. As far as the source current is concerned, these two currents tend to cancel. I_L is larger than I_C. Thus, the total current through the two reactive components is I_L minus I_C or 1 ampere minus 0.4 amperes equals 0.6 amperes. Because the reactive currents subtract, the formula for the total current can be written:

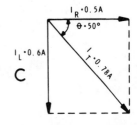

$$I_T = \sqrt{I_R^2 + (I_L - I_C)^2}$$

Figure 6-5
Parallel RLC Circuit

235

We will use this formula and solve for the total current in Figure 6-5A.

$$I_T = \sqrt{I_R^2 + (I_L - I_C)^2}$$

$$I_T = \sqrt{0.5^2 + (1 - 0.4)^2}$$

$$I_T = \sqrt{0.5^2 + (0.6)^2}$$

$$I_T = \sqrt{0.25 + 0.36}$$

$$I_T = \sqrt{0.61}$$

$$I_T = 0.78 \text{ A}$$

This is further illustrated by the vector shown in Figure 6-5C. The total current is shown as the vector sum of I_R and I_L. Of course, the value of I_L is the resultant current after the value of I_C is subtracted from the original value of I_L.

Once we know the total current in the circuit, we can compute other values. For example, the impedance of the circuit is:

$$Z = \frac{E}{I_T} = \frac{10 \text{ V}}{0.78 \text{ A}} = 12.8 \ \Omega$$

This is the impedance of the three components in parallel.

Also, the phase angle can be found by the formula:

$$\text{Tan } \theta = \frac{I_X}{I_R} = \frac{0.6}{0.5} = 1.2$$

Using a tangent table or a calculator, you will find that a tangent of 1.2 corresponds to an angle of about 50°. However, as shown in Figure 6-5C, θ is a negative angle because the circuit acts inductive. Consequently, the total current lags the applied voltage by about 50°.

RESONANCE

In earlier units you studied the inductive reactance (X_L) of coils and the capacitive reactance (X_C) of capacitors. Figure 6-6A shows a coil and a capacitor connected in parallel. If a voltage is applied between terminals A and B, the operation of this circuit will depend on the frequency of the applied voltage. If the voltage is dc, the capacitor will act like an open while the inductor will act like a short. That is, the X_C will be infinite while the X_L will be 0.

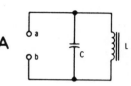

Now, assume that the applied voltage is not dc but is a very low frequency ac instead. In this case, the X_L will be very low and the X_C will be quite high. The exact values will depend on the values of C, L, and the applied frequency.

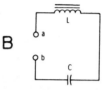

If the frequency is gradually increased, the X_L will gradually increase while the X_C will gradually decrease. As the frequency is increased further, a point is eventually reached at which the value of X_L is the same as the value of X_C. That is, for any combination of L and C, there is some frequency at which X_L equals X_C. This is true whether the two components are connected in parallel as shown in Figure 6-6A or in series as shown in Figure 6-6B. The condition at which X_L is equal to X_C is called resonance. Also, the frequency at which X_L is equal to X_C is called the resonant frequency and is abbreviated f_o.

Figure 6-6
LC Circuits

Recall that the formula for X_L is:

$$X_L = 2\pi\, fL$$

And, the formula for X_C is:

$$X_C = \frac{1}{2\pi\, fC}$$

Using these two formulas, we can derive an equation for the resonant frequency (f_o). By definition, at resonance, $X_L = X_C$. Therefore:

$$2\pi\, fL = \frac{1}{2\pi\, fC}$$

237

To solve for the frequency at which L and C are resonant, we can re-arrange this formula. First we multiply both sides of the equation by 2π fC. This gives us:

$$2\pi \text{ fL } (2\pi \text{ fC}) = \frac{1 \; (2\pi \text{ fC})}{2\pi \text{ fC}}$$

Or:

$$2\pi \text{ fL } (2\pi \text{ fC}) = 1$$

Simplifying we find that:

$$4\pi^2 \text{ f}^2 \text{ LC} = 1$$

Dividing both sides by $4\pi^2$LC we get:

$$\text{f}^2 = \frac{1}{4\pi^2 \text{ LC}}$$

Now, we take the square root of both sides:

$$\sqrt{\text{f}^2} = \sqrt{\frac{1}{4\pi^2 \text{ LC}}}$$

Of course, $\sqrt{\text{f}^2} = \text{f}$ and $\sqrt{\dfrac{1}{4\pi^2 \text{ LC}}}$ can be written $\dfrac{\sqrt{1}}{\sqrt{4\pi^2 \text{ LC}}}$.

Thus our equation becomes:

$$\text{f} = \frac{\sqrt{1}}{\sqrt{4\pi^2 \text{ LC}}}$$

The square root of 1 is 1 and the square root of $4\pi^2$ is 2π. Thus, our final equation is:

$$\text{f}_o = \frac{1}{2\pi \sqrt{\text{LC}}}$$

Or, dividing 2π into 1:

$$\text{f}_o = \frac{0.159}{\sqrt{\text{LC}}}$$

238

We can apply this formula to any LC circuit to find the frequency at which it is resonant. For example, assume that a 100 mH coil is connected in series with a .022 μfd capacitor. What is the resonant frequency?

$$L = 0.1 \text{ H}$$

$$C = .000\ 000\ 022 \text{ fd}$$

$$f_0 = \frac{0.159}{\sqrt{LC}}$$

$$f_0 = \frac{0.159}{\sqrt{0.1 \text{ H} \times 0.000\ 000\ 022 \text{ fd}}}$$

$$f_0 = \frac{0.159}{\sqrt{0.000\ 000\ 0022}}$$

$$f_0 = \frac{0.159}{0.000047}$$

$$f_0 = 3383 \text{ Hz or } 3.383 \text{ kHz}$$

Now try another example. A 5 μfd capacitor is in parallel with a 50 μH coil. What is the resonant frequency?

$$L = 0.000\ 05 \text{ H}$$

$$C = 0.000\ 005 \text{ fd}$$

$$f_0 = \frac{0.159}{\sqrt{0.000\ 05 \text{ H} \times 0.000\ 005 \text{ fd}}}$$

$$f_0 = \frac{0.159}{\sqrt{0.000\ 000\ 000\ 25}}$$

$$f_0 = \frac{0.159}{0.0000159}$$

$$f_0 = 10{,}000 \text{ Hz or } 10 \text{ kHz}$$

By rearranging the resonance formula in a different way, we can derive two more important equations. First we can derive an equation for finding the value of capacitance needed to resonate with a given value of inductance at a given frequency. As before:

$$X_L = X_C$$

$$2\pi \, fL = \frac{1}{2\pi \, fC}$$

$$4\pi^2 \, f^2 \, LC = 1$$

$$C = \frac{1}{4\pi^2 \, f^2 \, L}$$

Or, we can rearrange the equation differently to give:

$$L = \frac{1}{4\pi^2 \, f^2 \, C}$$

This equation can be used for finding the value of inductance needed to resonate with a given value of capacitance at a given frequency.

What value of capacitor must be connected across a 5 henry coil to make the circuit resonate at 60 Hz?

$$L = 5 \text{ H}$$

$$f = 60 \text{ Hz}$$

$$C = \frac{1}{4\pi^2 \, f^2 \, L}$$

$$C = \frac{1}{4 \, (3.14)^2 \, (60)^2 \, (5)}$$

$$C = \frac{1}{39.4 \, (3600) \, (5)}$$

$$C = \frac{1}{709\,200}$$

$$C = 0.0000014 \text{ fd}$$

$$C = 1.4 \; \mu\text{fd}$$

What value of coil must be connected in series with a 1 μfd capacitor in order that the circuit resonate at 5 kHz?

$$C = 0.000\ 001\ fd$$

$$f_0 = 5000\ Hz$$

$$L = \frac{1}{4\pi^2\ f^2\ C}$$

$$L = \frac{1}{4\ (3.14)^2\ (5000)^2\ (.000\ 001)}$$

$$L = \frac{1}{39.4\ (25,000,000)\ (.000\ 001)}$$

$$L = \frac{1}{985}$$

$$L = .00102\ H$$

$$L = 1.02\ mH$$

These examples show how we can find the resonant frequency, the value of L, or the value of C. Using these formulas, we can find any one of these variables, if the other two are known. These formulas work equally well for both series and parallel LC circuits. That is, for given values of L and C, the resonant frequency is the same regardless of how L and C are connected. Nevertheless, as you will see later, a series resonant circuit behaves very differently than a parallel resonant circuit.

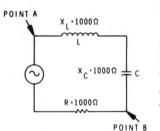

POINT A

$X_L \cdot 1000\Omega$

L

$X_C \cdot 1000\Omega$ C

$R \cdot 1000\Omega$

POINT B

Figure 6-7

At resonance $X_L = X_C$.

The previous section defined resonance and gave a formula for determining the resonant frequency of any LC circuit. In this section you will see that an LC circuit has characteristics at resonance which it does not have at any other frequency. These unusual characteristics make the resonant circuit extremely important.

Figure 6-7 shows a series circuit consisting of a capacitor, an inductor, and a resistor. The values of L and C are not given, nor is the frequency of the applied signal. Nevertheless, we know that the circuit is at the resonant condition because X_L is equal to X_C.

The applied ac signal forces current to flow through the series circuit. Because the components are in series, the same current flows through all components.

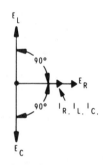

E_L

90°

E_R

90° $I_R \cdot I_L \cdot I_C$

E_C

Figure 6-8

Vector diagram of series resonant circuit.

As current flows through the circuit, a voltage is developed across each component. Because $R = X_L = X_C$, the voltages developed across each component are equal. That is, $E_R = E_L = E_C$. However, this is true only of the *magnitude* of the voltage. The phase of the voltage is different across each component. The voltage across R is in phase with the circuit current. However, the voltage across the coil leads the current by 90°. Also the voltage across the capacitor lags behind the current by 90°. The best way to visualize these phase relationships is with vector diagrams.

Figure 6-8 shows the vector diagram of the voltages and current. I_R, I_L, I_C, and E_R are all in phase with the applied current. Consequently, these are shown at 0°. E_L is drawn at +90° while E_C is drawn at −90°. Thus, E_L is 180° out of phase with E_C. This illustrates one of the most important characteristics of the series resonant circuit. At resonance, the voltages across the capacitor and coil are of equal magnitude but they are 180° out of phase. Consequently, E_L exactly cancels E_C. Thus, as far as the source is concerned, the sum of these two voltages is zero. This means that a voltmeter connected between points A and B in Figure 6-7 will read 0 volts.

242

If the combined voltage drop across X_L and X_C is 0, then the applied voltage must be developed across R. That is, if the source voltage is 10 volts, then the voltage across R will be 10 volts. A voltmeter placed across R will read 10 volts. However, since the same current must flow through C, a voltmeter placed across C will also read 10 volts. Moreover, a voltmeter placed across L will read 10 volts. At first this may seem to violate Kirchhoff's voltage law. It does not because the 10 volts across C is cancelled by the 10 volts across L. As Figure 6-8 shows, the vector sum of E_L and E_C is 0 volts. This is the first of several strange occurances which take place in series resonant circuits.

Another strange thing is the apparent disappearance of the coil and the capacitor at resonance. Since the combination of L and C produces no voltage drop, their total reactance or impedance must be zero. That is, the source sees the LC combination as a perfect conductor having 0 ohms of impedance. Thus, the only opposition to current flow in the circuit is the resistance of R. The source sees no capacitance and no inductance, only resistance.

This can be easily proven by applying the impedance formula discussed earlier. Recall that:

$$Z = \sqrt{R^2 + (X_L - X_C)^2}$$

When $X_L = X_C$, $X_L - X_C = 0$. Therefore:

$$Z = \sqrt{R^2 + 0}$$

$$Z = \sqrt{R^2}$$

$$Z = R$$

Thus at resonance, the total impedance is simply the value of R. This means that the current and voltage as seen by the source are in phase. Consequently, the power factor is 1.

It is important to emphasize again that these conditions occur only at resonance. When the applied frequency is above or below resonance, X_L does not exactly equal X_C. Consequently, the voltage drops across L and C do not completely cancel. In this case, there is some resultant value of reactance.

243

The third strange thing which happens in series resonant circuits is the most mysterious of all. It is not evident in Figure 6-7 because the value of R is equal to the value of X_L and X_C. However, look at the circuit shown in Figure 6-9. Here the value of R is much less than that of X_L or X_C. Nevertheless, the circuit is still at resonance because X_L equals X_C.

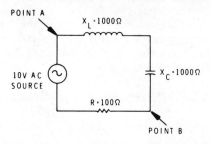

Figure 6-9
R is less than X_L or X_C.

You have seen that the current in a resonant circuit is determined solely by the applied voltage and the value of the series resistance. Thus, in this circuit, $I = \dfrac{E}{R} = \dfrac{10 \text{ V}}{100 \text{ }\Omega}$ = 0.1 ampere. This same current flows through C and L. Therefore, the voltage drop across C must be $E_C = I\,(X_C)$ = 0.1 A × 1000 Ω = 100 V. Also, the voltage drop across L must be $E_L = I$ (X_L) = 0.1 A × 1000 Ω = 100 V. Notice that the voltage across L or C is actually ten times higher than the voltage being applied to the circuit.

Once again, this may seem to violate Kirchhoff's voltage law. However, it does not since the 100 volts across L is exactly cancelled by the 100 volts across C. A voltmeter connected from point A to point B will still read 0 volts. But, if the meter is connected across L or C separately, it will read 100 volts.

This ability to produce a voltage higher than the applied voltage is one of the most remarkable characteristics of the series resonant circuit. This is possible because of the ability of the coil and the capacitor to store energy.

244

It occurs at resonance any time that the value of R is lower than the value of X_L or X_C. The lower the value of the resistance as compared to the reactance, the higher the voltage across the reactance will be. If all series resistance could be eliminated entirely, the current in the circuit would rise to an infinitely high value. The voltage across the coil and the capacitor would also become infinitely high.

In practice, of course, some series resistance will always be present. The ac source will always have some value of series resistance as will the connecting wires. However, the largest source of series resistance is generally the coil. Most coils are wound from lengths of very small wire. Thus, they have a relatively large value of series resistance. This resistance tends to keep the resonant current down even if no separate resistor is used.

To summarize, the series resonant circuit has several important characteristics. These are listed below:

1. The impedance across the circuit is low and is equal to the series resistance.

2. The current flow is high and is limited only by the series resistance.

3. The applied voltage is dropped by the series resistance.

4. The voltage across the coil or capacitor is equal to the current times the reactance. This voltage may be higher than the applied voltage.

5. The circuit acts resistive. The source current and voltage are in phase and the power factor is 1.

Q AND BANDWIDTH
IN SERIES RESONANT CIRCUITS

The series resonant circuit has two characteristics which we have not yet discussed. These are called Q and *bandwidth*. These two characteristics are defined and explained in this section. Q was discussed earlier as a figure of quality for coils. However, the Q used in connection with resonant circuits has additional aspects which must be understood. Bandwidth was also mentioned earlier in connection with filters. Here, we will discuss bandwidth in more detail.

Q In Series Resonant Circuits

One of the most important characteristics of a resonant circuit is its Q. Other names for Q include: quality figure, figure of merit, and magnification factor. The Q factor is defined as the ratio of the reactance at resonance to the series ac resistance. That is,

$$Q = \frac{X_L}{R}$$

Or, since at resonance $X_L = X_C$,

$$Q = \frac{X_C}{R}$$

Normally though, we express the reactance in terms of X_L.

In a resonant circuit in which $X_L = X_C = 1000 \ \Omega$ and the series ac resistance is 100 Ω; the Q factor is 10 because:

$$Q = \frac{X_L}{R} = \frac{1000 \ \Omega}{100 \ \Omega} = 10$$

Since Q is the ratio of reactance to resistance, the ohms in each term cancel. Thus, Q is simply a number without any associated units.

In the series resonant circuit, Q is the magnification factor that determines how much the voltage across L or C is increased above the applied
voltage. For example, if a 1 V peak-to-peak ac signal is applied across a
series resonant circuit with a Q of 10, the voltage across L or C at
resonance will be 10 V peak-to-peak. Thus, the applied voltage (E_{in}) is
magnified by the Q factor. Expressed as an equation:

$$E_L = Q \times E_{in}$$

And:

$$E_C = Q \times E_{in}$$

When the applied voltage (E_{in}) and the voltage across L or C (E_L or E_C) are
known, the Q factor can be calculated using the formula:

$$Q = \frac{E_L}{E_{in}} \quad \text{or} \quad Q = \frac{E_C}{E_{in}}$$

Q can be determined by measuring E_{in} and E_L or E_C and using the above
formula. For example, E_{in} is measured with an ac voltmeter and found to
be 0.1 volt. E_L is measured and found to be 15 volts. The Q of the circuit
must be:

$$Q = \frac{E_L}{E_{in}} = \frac{15 \text{ V}}{0.1 \text{ V}} = 150.$$

This method of determining Q generally gives more accurate results than
the X_L/R method. The reason for this is that the ac resistance of the circuit
is difficult to determine.

Generally, the largest single factor that makes up the series resistance is
the ac resistance of the coil. This ac resistance can be much higher than
the dc resistance measured with an ohmmeter. This makes it difficult to
measure the ac resistance of the coil directly.

You will recall that a coil has a Q factor of its own. If the only series resistance in a series resonant circuit is that of the coil, the Q of the circuit will be the same as that of the coil. The Q of the coil is the highest possible value of Q that a resonant circuit can have. If additional series resistance is added, the Q of the circuit will be less than the Q of the coil.

While it is difficult to measure directly the ac resistance of a series resonant circuit, the value can be computed from known or measured circuit quantities. For example, a series resonant circuit develops 10 volts across a 1 henry coil at 100 Hz with a 0.1 volt ac input. Find the Q and the ac resistance.

$$Q = \frac{E_L}{E_{in}} = \frac{10 \text{ V}}{0.1 \text{ V}} = 100$$

We know that $Q = \frac{X_L}{R}$. Consequently: $R = \frac{X_L}{Q}$. We have seen that Q is 100. Thus, if we find X_L, we can determine R.

$$X_L = 2\pi \, fL$$

$$X_L = 6.28 \, (100 \text{ Hz}) \, (1\text{H})$$

$$X_L = 628 \, \Omega$$

Therefore:

$$R = \frac{X_L}{Q}$$

$$R = \frac{628\ \Omega}{100}$$

$$R = 6.28\ \Omega$$

This shows some of the reasons that Q is important. However, Q is also important in determining the bandwidth of a resonant circuit.

Bandwidth and Q

Resonant circuits are selective. They respond more readily to their resonant frequency (f_o) than to other frequencies. While the resonance effects are greatest at f_o, these same effects exist to a smaller extent at frequencies slightly above and below f_o. Thus, a resonant circuit actually responds to a band of frequencies. The width of this band of frequencies is called the *bandwidth* of the resonant circuit.

Measuring Bandwidth

Figure 6-10 illustrates how bandwidth is measured. This graph shows the current passed by a series resonant circuit at various frequencies below, at, and above the resonant frequency. Naturally, maximum current flows at the resonant frequency. In this example, f_o is 1000 Hz and the maximum current is 10 mA.

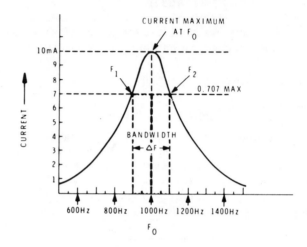

Figure 6-10
Bandwidth is measured between
half-power points

The bandwidth of the series resonant circuit is generally considered to include that group of frequencies with a response 70.7 % of maximum. In our example, frequencies which produce a current of 7.07 mA or higher are considered within the bandwidth. This band of frequencies extends from f_1 (900 Hz) to f_2 (1100 Hz).

The *bandwidth* is defined as the width of this band of frequencies. Consequently the bandwidth (BW) is $f_2 - f_1$, or 1100 Hz − 900 Hz = 200 Hz. That is, the bandwidth is the width of the band of frequencies which produce a response of 70.7% of maximum current.

250

Half-Power Points. You may wonder why the 70.7% points were chosen to indicate the bandwidth. Actually this is a very convenient point to use because it represents the point at which the power in the circuit is exactly one-half the maximum value. Thus, the points marked f_1 and f_2 in Figure 6-10 are referred to as *half-power points*.

An example demonstrates that the power in a circuit drops to one-half when the current drops to 70.7%. Consider a circuit in which the resistance is 2000 ohms and the current is 10 mA. The power is: $P = I^2R = 0.01^2 \times 2000 = 0.0001 \times 2000 = 0.2$ W. Now, assume that the current drops to 70.7% of maximum or to 7.07 mA. The power drops to: $P = I^2R = 0.00707^2 \times 2000 = 0.00005 \times 2000 = 0.1$ W. This is one-half the previous power. Thus, reducing the current to 70.7% reduces the power to 50%. For convenience then, the bandwidth is measured between half-power points.

Bandwidth Equals f_o/Q. With the bandwidth measured between the half-power points, an interesting relationship exists between the bandwidth, the resonant frequency, and the value of Q. This relationship is expressed by the equation:

$$BW = \frac{f_o}{Q}$$

This states that the bandwidth is equal to the resonant frequency divided by the Q.

For the example shown in Figure 6-10, the Q is 5 because:

$$BW = \frac{f_o}{Q}$$

$$BW = \frac{1000 \text{ Hz}}{5}$$

$$BW = 200 \text{ Hz}$$

The equation states that the bandwidth is directly proportional to the resonant frequency but inversely proportional to the value of Q.

The curves shown in Figure 6-11 illustrate that the bandwidth increases as the value of Q decreases. When the value of Q is high, the current in the circuit is relatively high. The resonant circuit responds to a very narrow band of frequencies. The resonant frequency is 100 kHz. Thus, if the Q is 50, the bandwidth is:

$$BW = \frac{f_o}{Q} = \frac{100 \text{ kHz}}{50} = 2 \text{ kHz}.$$

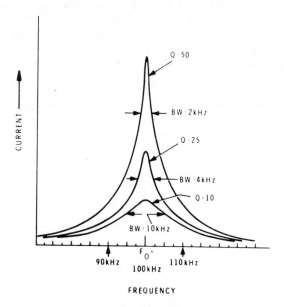

Figure 6-11

Bandwidth increases as Q decreases.

Notice what happens to the curve when the Q is reduced to 25. The current is lower and the curve is somewhat broader. The bandwidth is increased to: $BW = \dfrac{f_o}{Q} = \dfrac{100 \text{ kHz}}{25} = 4 \text{ kHz}.$

252

When the Q is reduced to 10, an even broader response curve results. The current is relatively low and the bandwidth is:

$$BW = \frac{f_o}{Q} = \frac{100 \text{ kHz}}{10} = 10 \text{ kHz}.$$

Figure 6-12 shows three circuits which produce curves like those shown in Figure 6-11. The three circuits are identical except for the value of the resistance. The value of R determines the Q of the circuit. This, in turn, determines the bandwidth. Notice that the value of R does not affect the resonant frequency, only the Q and bandwidth. As the value of R increases, the Q decreases and the bandwidth increases.

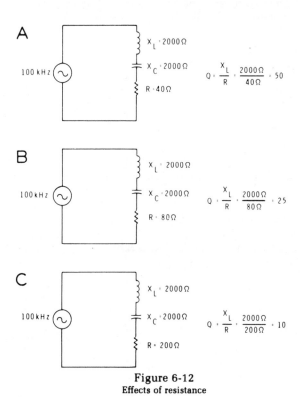

Figure 6-12
Effects of resistance

253

PARALLEL RESONANCE

Up to now we have considered only those resonant circuits in which the capacitor is in series with the inductor. In this section, we will consider another type of resonant circuit called a parallel resonant circuit. When the capacitor is placed in parallel with the inductor, the characteristics of the resonant circuit change completely.

Ideal Circuit

A parallel resonant circuit is shown in Figure 6-13. We know that the circuit is in the resonant condition because X_L is equal to X_c. Thus, the applied ac signal is at the proper frequency to cause the circuit to resonate.

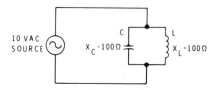

Figure 6-13
Parallel resonant circuit.

To simplify the explanation of this circuit, initially assume that L and C are ideal components so that there is no resistance in the circuit. Of course, in practical circuits there is always some resistance and we will consider its effects later. But for now, we will discuss what would happen in an ideal parallel resonant circuit.

If the capacitor is temporarily disconnected from the circuit as shown in Figure 6-14A, the current through the coil can be determined by Ohm's law:

$$I_L = \frac{E}{X_L} = \frac{10\ V}{100\ \Omega} = 0.1\ A$$

This current must be supplied by the ac source. Since L is a pure inductor, I_L must lag the applied voltage by 90°.

If the capacitor is reconnected and the inductor is disconnected as shown in Figure 6-14B, the current through the capacitor can be determined by Ohm's law:

$$I_C = \frac{E}{X_C} = \frac{10\ V}{100\ \Omega} = 0.1\ A$$

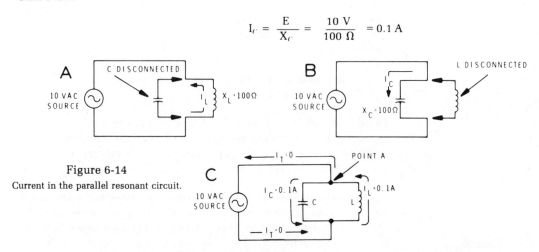

Figure 6-14
Current in the parallel resonant circuit.

However, you will recall that in a capacitor the current leads the voltage by 90°. Thus, I_C must lead the applied voltage by 90°. Now consider the operation of the complete circuit. If I_C leads the applied voltage by 90° and I_L lags the applied voltage by 90°, then I_C must be 180° out of phase with I_L. This means that when the current is flowing in one direction through L, an equal current must be flowing in the opposite direction through C.

At the instant when 0.1A is flowing up through L, exactly 0.1A must be flowing down through C. Now if we apply Kirchhoff's current law to point A in Figure 6-14C, we discover that there is no current flowing into or out of the source. That is, the same current that flows up through L also flows down through C and no current flows to or from the source. On the next alternation, the current flows up through C and down through L but still no current flows in the external circuit. The current simply oscillates back and forth between the capacitor and the coil.

255

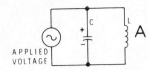
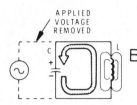

APPLIED
VOLTAGE

In the ideal parallel resonant circuit, the source voltage is required only to start the oscillation. Once started, the source can be disconnected and the oscillations will continue indefinitely. As mentioned earlier, this is true only if there are no losses in the circuit.

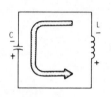

APPLIED
VOLTAGE
REMOVED

Now we will consider how the parallel resonant circuit appears to the ac source. The ac voltage is applied across the LC combination, and yet no current flows to the source. Consequently, as far as the source in concerned, the circuit appears to be open. That is, it appears to have infinite impedance.

Flywheel Effect

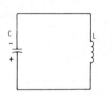

The ability of a parallel resonant circuit to sustain oscillation after the source voltage is removed is called the flywheel effect. It gets this name because the action is similar to that of a mechanical flywheel. Once the mechanical flywheel is started it tends to keep going until stopped by friction or some outside force. The flywheel effect of the parallel resonant circuit is illustrated in Figure 6-15.

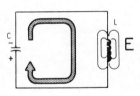

Initially, energy is supplied to the circuit by an ac source. Once started, the energy is alternately stored by the capacitor and then by the coil. We will pick up the action at the point where C is fully charged as shown in Figure 6-15A.

The applied voltage can be removed since it has supplied the necessary starting energy. As shown in Figure 6-15B, the capacitor begins to discharge through L. As current flows through L, a magnetic field builds up around the inductor.

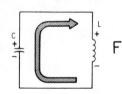

When C is discharged, the current through L tends to stop. Consequently, the magnetic field around L collapses inducing an emf with the polarity shown in Figure 6-15C. This keeps the current flowing in the same direction and charges C to the polarity shown.

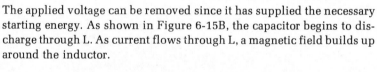

When the magnetic field has collapsed, the condition shown in Figure 6-15D exists. Here the capacitor is again fully charged. At the next instant, the capacitor begins to discharge. This time, the current flows in the opposite direction through the coil as shown in Figure 6-15E. This causes a magnetic field of the opposite polarity to build up around L.

Figure 6-15
The flywheel effect.

When C is again discharged, the current through L tries to stop and the magnetic field collapses. As shown in Figure 6-15F, and emf is induced which tends to keep current flowing in the same direction. Thus, C is again charged to its initial polarity.

At this point the cycle repeats itself. As you can see, the energy is simply interchanged between the capacitor and the coil. Initially, the energy is stored as an electrostatic field in the capacitor. Then, it is stored as a magnetic field around the inductor. Since neither the capacitor nor the coil dissipates energy, the oscillations would continue indefinitely if there were no losses in the circuit.

Because a circuit of this type can store energy, it is commonly called a *tank circuit*.

Practical Tank Circuits

The ideal tank circuit is one which has no resistance and no losses of any kind. Unfortunately, such a tank circuit does not exist. The capacitor, the coil, and the interconnecting wires all have resistance. Normally, though, only the resistance of the coil is high enough to be of importance. Thus, in reality, practical tank circuits must be analyzed as if there was a resistor in series with the inductor as shown in Figure 6-16. R represents the resistance of the coil.

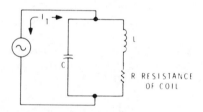

Figure 6-16
Practical tank circuit

257

Unlike reactance, resistance dissipates power. As the current oscillates between the coil and the capacitor, the resistor dissipates some of the power in the form of heat. Consequently, some of the energy is removed from the tank circuit during each cycle. For this reason, in a practical tank circuit, the oscillations will quickly die out if the ac source is disconnected. Figure 6-17 shows how the voltage waveform across the tank circuit would look. Each cycle gets progressively weaker as the resistance gradually dissipates the energy stored in the circuit. The waveform produced is called a damped sine wave.

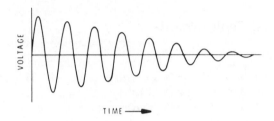

Figure 6-17
Damped sine wave.

Because of the power loss in the circuit, power must be supplied from the source to make up for the loss. Thus, the ac source provides just enough power to make up for that lost by the resistance. The result is that some current will flow from the ac source to the tank circuit. If the resistance value in the tank is high, more power will be dissipated, and the current from the source will be higher.

This may seem to contradict Ohm's law since current is inversely proportional to resistance. The resistance does limit the circulating current within the tank. However, in doing so, it consumes power. This power must be supplied by the ac source. Therefore, the current from the source must increase.

Q in Parallel Resonant Circuits

In the series resonant circuit we found Q by dividing the applied voltage (E_{in}) into either E_C or E_L. This will not work in the parallel resonant circuit since E_{in} is applied directly across both C and L. In the parallel resonant circuit, we are concerned with current rather than with voltage. Thus, in the parallel resonant circuit, Q can be determined by dividing the source current into the tank current. Recall that in a good tank circuit the source current is quite low while the circulating current can be very high. Therefore,

$$Q = \frac{I_{TANK}}{I_{SOURCE}}$$

If the source current is 1 mA and the tank current is 100 mA then the Q is:

$$Q = \frac{I_{TANK}}{I_{SOURCE}} = \frac{100\ \text{mA}}{1\ \text{mA}} = 100$$

As with the series resonant circuit, Q can also be expressed as the ratio of X_L (or X_C) to R. That is: $Q = \dfrac{X_L}{R}$. R is the total ac resistance within the tank. It can be somewhat higher than the value of R measured with an ohmmeter.

In the parallel resonant circuit Q can be thought of as a magnification factor. However, in this case it is not the voltage that is being magnified but the impedance. Because the source current is minimum at resonance, the impedance of the tank circuit is maximum at resonance. The impedance of the tank is equal to the reactance of L or C times the Q. That is: $Z_{TANK} = X_L$ (Q) or $Z_{TANK} = X_C$ (Q). If the value of X_L (or X_C) at resonance is 1000 ohms and the Q is 100, then the impedance of the tank is:

$$Z_{TANK} = X_L\ (Q)$$

$$Z_{TANK} = 1000\ \Omega \times 100$$

$$Z_{TANK} = 100{,}000\ \Omega \text{ or } 100\ \text{k}\Omega$$

By transposing the above equation, we can develop another useful equation. That is:

$$Q = \frac{Z_{TANK}}{X_L}$$

Thus we have three equations for Q. These are:

$$Q = \frac{I_{TANK}}{I_{SOURCE}} \qquad Q = \frac{X_L}{R} \qquad Q = \frac{Z_{TANK}}{X_L}$$

Of these equations, the last one is the most useful for determining Q. It is difficult to use the first equation because it is hard to measure the ac current. The second equation also presents problems because it is hard to determine the total ac resistance.

To use the last equation, all we need is the values of X_L and Z_{TANK}. X_L can be easily computed if the value of L is known. Furthermore, the impedance of the tank can be easily determined by the method shown in Figure 6-18.

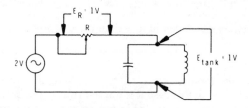

Figure 6-18

When $E_R = E_{TANK}$, $Z_{TANK} = R$

Here, a variable resistor is placed in series with the tank circuit. At the resonant frequency, the resistor is adjusted until the voltage across the resistor is equal to the voltage across the tank. That is: $E_R = E_{TANK} = 1/2\ E_{in}$. At this point, R drops the same amount of voltage as the tank. Thus, R must be equal to the impedance of the tank. The circuit is then disassembled and the value of R is measured with an ohmmeter. This tells us the value of Z_{TANK}. This value can then be used in the equation to determine the Q of the circuit.

Bandwidth in Parallel Resonant Circuits

Like the series resonant circuit, the parallel tank circuit responds to a band of frequencies rather than a single frequency. Figure 6-19 shows a typical response curve. Notice that this curve has the same shape as the response curve shown earlier for the series resonant circuit. However, if you examine Figure 6-19 more closely you will see that it shows the impedance of the circuit rather than the current through the circuit.

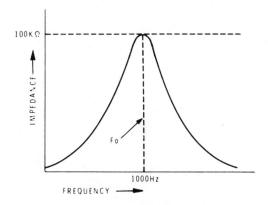

Figure 6-19

Response curve of parallel resonant circuit. (Impedance
vs. Frequency)

At the resonant frequency, the impedance is maximum. Below resonance, the coil offers a low reactance and the impedance falls off. Above
resonance, the capacitor offers a low reactance and the impedance again
falls off.

Because the source or line current is inversely proportional to the impedance, the current response curve has the shape shown in Figure 6-20.
Notice that the line current decreases as the resonant frequency is ap-
proached.

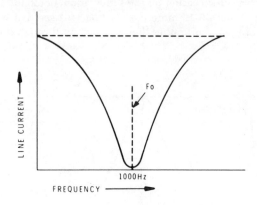

Figure 6-20

Response curve of parallel re-

sonant circuit. (Current vs.

Frequency)

As with the series resonant circuit, the width of the band of frequencies to
which the circuit responds is determined by the Q of the circuit. The
bandwidth is determined by the formula: $BW = \dfrac{f_o}{Q}$. If a parallel resonant circuit has a resonant frequency of 1000 Hz and a Q of 20, the
bandwidth will be:

$$BW = \frac{f_o}{Q} = \frac{1000 \text{ Hz}}{20} = 50 \text{ Hz}$$

Thus, it would respond to the band of frequencies between 975 Hz and 1025 Hz.

Often a parallel resonant circuit will be more selective than we would like. That is, it responds only to a very narrow band of frequencies. In these cases, we can increase the bandwidth by connecting a relatively small value resistor across the tank circuit as shown in Figure 6-21A. The resistor provides an alternate path for line current. Thus, the line current increases. Recall that $Q = \dfrac{I_{TANK}}{I_{LINE}}$. For this reason, Q is inversely proportional to the line current. If the line current increases then Q must decrease. However, the formula $BW = \dfrac{f_0}{Q}$ shows that the bandwidth is inversely proportional to Q. Thus, if Q decreases, the bandwidth increases. Therefore, connecting a resistor across the tank circuit increases the bandwidth. This is often called *loading* the tank circuit.

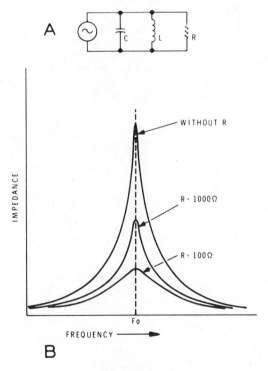

Figure 6-21
Bandwidth may be increased
by loading the tank circuit

Figure 6-21B shows the effect that the resistor can have. Without the resistor, the impedance is extremely high and the circuit responds only to a very narrow band around the resonant point. When a 1000 ohm resistor is added, the additional parallel path for current reduces the impedance and broadens the bandwidth. A 100 ohm resistor reduces the impedance still further and stretches the bandwidth even more.

Distributed Capacitance and Self Resonance of Coils

As mentioned earlier, every coil has a certain value of distributed capacitance. That is, the coil acts as if a small value capacitor is connected in parallel. At some frequency, the coil and this value of distributed capacitance form a parallel resonant circuit. At this frequency, the coil is said to be self-resonant. Thus, a single coil can have the characteristics of a parallel resonant circuit at its self-resonant frequency.

LC FILTERS

In an earlier unit you saw that RC and RL circuits pass some frequencies more easily than others. When an RC or RL circuit is especially designed to be frequency selective, the resulting circuit is called a filter.

In this unit, you have seen that LC circuits are frequency selective. Therefore LC circuits can make good filters.

Types of Filters

There are several types of filters used in electronics. Generally, a basic filter circuit can be placed in one of four categories.

A *band-pass* filter is designed to pass a narrow band of frequencies while rejecting both higher and lower frequencies. Such filters are used in radios and TV receivers to pass the frequencies of the desired station while blocking the frequencies of all other stations.

A *band-stop* filter is designed to pass all frequencies except a narrow band. This type of filter can be used to clip out an annoying frequency without interfering with wanted frequencies.

A *low-pass* filter is used to attenuate all frequencies above a certain *cutoff* frequency. Thus, it passes low frequencies but blocks high frequencies.

A *high-pass* filter has the opposite characteristics. It passes signals above the *cutoff* frequency but blocks those below.

All four types of filters are commonly used. The frequencies which are blocked or passed depend on the component values used and how the components are arranged. Next, we will look at some typical filters starting with the band-pass filter.

Band-Pass Filter

A very simple band-pass filter is shown in Figure 6-22A. The filter is the series resonant circuit composed of L and C. R_L is the load to which the voltage is applied. At the resonant frequency, the series resonant circuit has a very low impedance. Thus, it drops very little of the applied voltage, E_{in}. Most of the voltage is developed across R_L. Consequently, E_{out} is high at the resonant frequency.

265

Below the resonant frequency, the X_C of the capacitor is higher than the resistance of R_L. Consequently, most of E_{in} is dropped by C. This leaves only a small voltage across R_L. Thus, E_{out} is a low voltage.

A

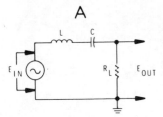

Above the resonant frequency, the X_L of the coil is higher than R_L. Therefore, most of the voltage is dropped across the coil and E_{out} is again a low voltage.

B

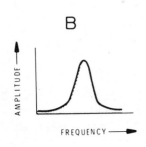

Figure 6-22B shows how the circuit responds to a band of frequencies. At the resonant frequency of L and C, E_{out} is quite high. Above and below resonance, E_{out} drops off quickly to a low voltage.

Figure 6-22C shows that the parallel resonant circuit can also be used as a band-pass filter. Whereas the series resonant circuit is placed in series with the output, the parallel resonant circuit is placed across the output. The reason for this becomes evident if we remember the characteristics of the parallel resonant circuit.

C

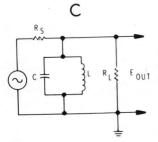

At resonance the impedance of the tank circuit is extremely high. Consequently, very little current flows through the tank circuit and most of the current flows through R_L. The current through R_L is maximum at resonance.

Figure 6-22
Band-pass filter.

Below resonance, the X_L of the coil is quite small compared to the value of R_L. Thus, most current flows through L and very little flows through R_L. Above resonance, most of the current flows through the capacitor leaving little current for R_L. That is, above and below resonance, R_L is partially shorted out by the low impedance of the tank. Thus, most of the applied voltage is dropped across R_S. However, at resonance, the impedance of the tank is high and R_L is no longer shorted.

Figure 6-23A shows another type of band-pass filter. This one uses a transformer. Recall that the windings of a transformer have an inductance value like any other coil. Therefore, a capacitor can be connected across one of the windings to form a parallel resonant circuit.

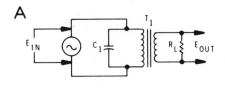

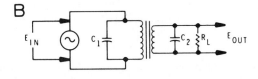

Figure 6-23
A tuned transformer can act as a band-
pass filter.

In Figure 6-23A, C_1 is connected across the primary of the transformer. This causes the transformer to respond much more readily to the resonant frequency than to other frequencies. Recall that at resonance the circulating current within the tank is at its maximum value. This heavy current in the primary of the transformer develops a strong magnetic field. Consequently, maximum voltage is coupled to the secondary at the resonant frequency. Frequently both the primary and the secondary will be tuned as shown in Figure 6-23B. The bandwidth of such a circuit depends mainly on three factors: the Q of the tuned primary circuit, the Q of the tuned secondary circuit, and the coefficient of coupling. When the coefficient of coupling is close to 1, the bandwidth will be extremely broad. However, when the coefficient of coupling is very low, the bandwidth will be quite narrow.

Band-Stop Filter

The response of the band-stop filter is opposite that of the band-pass filter. That is, the band-stop filter stops, attenuates, or rejects the frequency to which it is tuned.

Figure 6-24A shows a simple band-stop filter. Here L and C form a parallel resonant circuit which is in series with the load (R_L). At resonance, the impedance of the tank circuit is much higher than that of R_L. Consequently, most of E_{in} is dropped across the tank and very little voltage is available at the load. Above and below resonance, the resistance of R_L is higher than the impedance of the tank. Therefore, most of E_{in} is developed across R_L. Figure 6-24B shows the response of the circuit. The sharpness of the curve is determined by the Q of the resonant circuit.

Another circuit that produces about the same response is shown in Figure 6-24C. Here a series resonant circuit is connected across the load. At resonance, the series resonant circuit offers a very low impedance to current flow. This shorts most of the current around the load. Most of the applied voltage is dropped across R_S. Above and below resonance, the impedance of the filter is much higher and R_L is no longer shorted out.

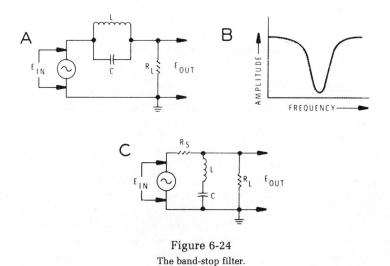

Figure 6-24
The band-stop filter.

268

Low-Pass Filter

A low-pass filter passes all frequencies below a certain cutoff frequency. A simple low-pass filter is shown in Figure 6-25A. At low frequencies X_L is lower than the resistance of R_L. Thus, most of E_{in} is developed across R_L. Furthermore, the X_c of the capacitor is high at low frequencies. Thus, most of the current flows through R_L. As you can see, E_{out} is quite high at low frequencies.

At high frequencies the situation reverses, the X_L of the coil increases dropping most of applied voltage. Only a slight voltage is developed across R_L. Furthermore, the X_c of the capacitor decreases so that most of the current is shunted around R_L. Hence, the filter effectively blocks high frequency signals. The response of the filter is shown in Figure 6-25B.

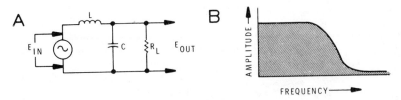

Figure 6-25
Low-pass filter.

High-Pass Filter

The high-pass filter passes all frequencies above a certain cutoff frequency. Figure 6-26 shows the high pass filter and its response curve. At high frequencies X_c is low and X_L is high. Thus, the capacitor and coil have little effect. Most of E_{in} is developed across R_L. At low frequencies, X_c is high and X_L is low. Thus, the high value of X_c drops most of the applied voltage while the low value of X_L tends to short R_L. Thus, the circuit passes high frequencies but blocks lower frequencies.

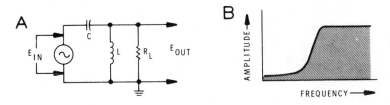

Figure 6-26
High-pass filter.

269

88. (capacitor, coil) Figure 6-27 shows the response curves of the four basic types of filters. Match the following:

1. band-pass   A. Figure 6-27A
2. band-stop   B. Figure 6-27B
3. high-pass   C. Figure 6-27C
4. low-pass   D. Figure 6-27D

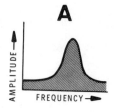
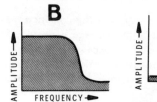
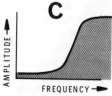
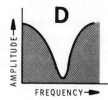

Figure 6-27

Curves for frame 88.

89. (1-A, 2-D, 3-C, 4-B) Figure 6-28 shows six types of filters. Match the following:

A. Figure 6-28A   1. band-pass
B. Figure 6-28B   2. band-stop
C. Figure 6-28C   3. high-pass
D. Figure 6-28D   4. low-pass
E. Figure 6-28E
F. Figure 6-28F

(A-4, B-1, C-1, D-3, E-2, F-2)

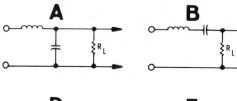
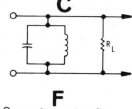
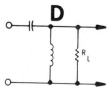
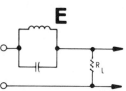
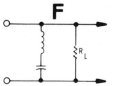

Figure 6-28

Circuits for Frame 89.

270

SUMMARY

In a series RLC circuit, the total impedance is found by using the formula:

$$Z = \sqrt{R^2 + (X_L - X_C)^2}$$

Or, if X_C is larger than X_L, the formula becomes:

$$Z = \sqrt{R^2 + (X_C - X_L)^2}$$

With parallel RLC circuits, it is easier to work with currents. The formula for finding the total current is:

$$I_T = \sqrt{I_R^2 + (I_L - I_C)^2}$$

Or, if I_C is larger than I_L, the formula is:

$$I_T = \sqrt{I_R^2 + (I_C - I_L)^2}$$

For any LC circuit, there is a frequency at which X_L is equal to X_C. This frequency is called the resonant frequency (f_o). The resonant frequency can be determined by the formula:

$$f_o = \frac{.159}{\sqrt{LC}}$$

271

Some of the characteristics of a series resonant circuit are quite different from those of parallel resonant circuit. The table in Figure 6-29 summarizes the differences and similarities. The conditions described are generally true for high Q resonant circuits.

SERIES RESONANT CIRCUIT	PARALLEL RESONANT CIRCUIT
Current maximum at resonance.	Line current minimum at resonance.
Impedance minimum at resonance.	Impedance maximum at resonance.
$Q = \dfrac{X_L}{R}$, $Q = \dfrac{E_C}{E_{in}}$, $Q = \dfrac{E_L}{E_{in}}$	$Q = \dfrac{X_L}{R}$, $Q = \dfrac{Z_{tank}}{X_L}$, $Q = \dfrac{I_{tank}}{I_{source}}$
Acts purely resistive at f_0.	Acts purely resistive at f_0.
At resonance, the source current and voltage are in phase.	At resonance, the source current and voltage are in phase.
Below resonance, the circuit acts capacitively.	Below resonance, the circuit acts inductively.
Above resonance, the circuit acts inductively.	Above resonance, the circuit acts capacitively.

Figure 6-29

Comparison of series and parallel resonant circuits.

272

One practical application of tuned circuits is the filter. Figure 6-30 summarizes the six basic types of filters discussed in this unit.

TYPE	TYPICAL CIRCUIT	RESPONSE CURVE	COMMENTS
BAND-PASS	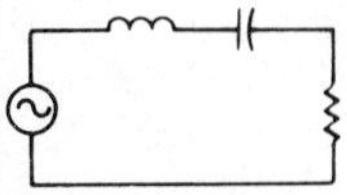	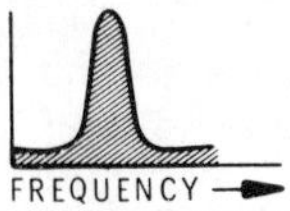	PASSES A BAND OF FREQUENCIES AROUND F_0.
BAND-PASS	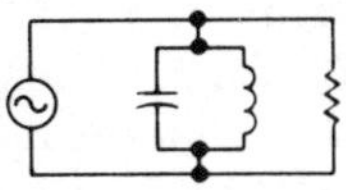	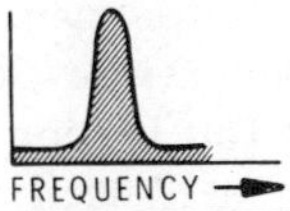	PASSES A BAND OF FREQUENCIES AROUND F_0.
BAND-STOP	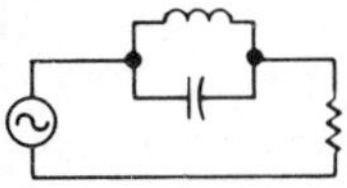	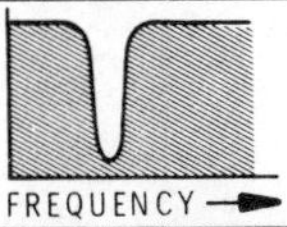	BLOCKS OR ATTENUATES A BAND OF FREQUENCIES AROUND F_0.
BAND-STOP	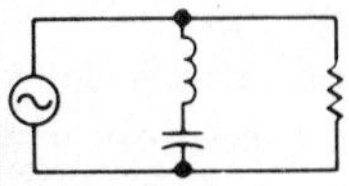		BLOCKS OR ATTENUATES A BAND OF FREQUENCIES AROUND F_0.
HIGH-PASS	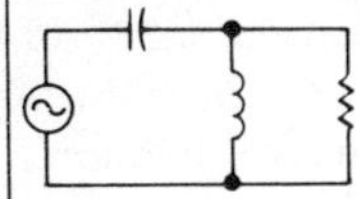	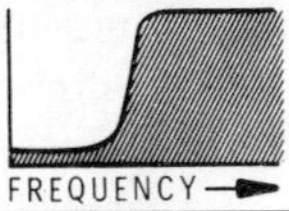	BLOCKS OR ATTENUATES FREQUENCIES BELOW A CERTAIN CUTOFF FREQUENCY.
LOW-PASS	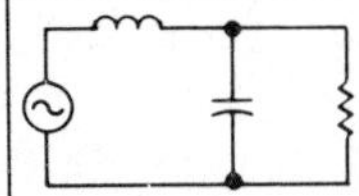	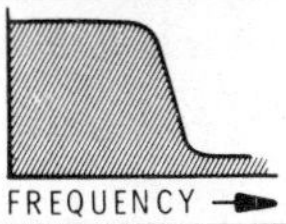	BLOCKS OR ATTENUATES FREQUENCIES ABOVE A CERTAIN CUTOFF FREQUENCY.

Figure 6-30
Comparison of filter types.

273

Appendix A

RESONANCE NOMOGRAPH

Figure A-1 is a nomograph which can be used for solving resonance problems. The accuracy of this graph is about 95% for most problems. It can be used for solving three types of resonance problems.

It can be used to determine the resonant frequency of a series or parallel resonant circuit when the inductance and capacitance values are known. For example, if a 100 mH coil is connected in series with a .022 μfd capacitor, what is the resonant frequency? A straight line is drawn from 0.022 μfd in the left column to 100 mH in the right column. The resonant frequency is read from the center column. Figure A-1 shows a resonant frequency of about 3.4 kHz. The exact value should be 3.383 kHz. So as you can see, the graph is fairly accurate.

The graph can also be used to determine the inductance when the capacitance and resonant frequency are known. For example, what value of inductance will resonant at 3.4 kHz when connected in parallel with a .022 μfd capacitor? A straight line is drawn from 0.22 μfd on the left column through 3.4 kHz in the center column. The inductance is read from the point at which this line crosses the right column.

Finally, the graph can be used to find the capacitance when the resonant frequency and the inductance are known. A straight line is drawn from the two known quantities. The capacitance is read from the point at which this line crosses the left column.

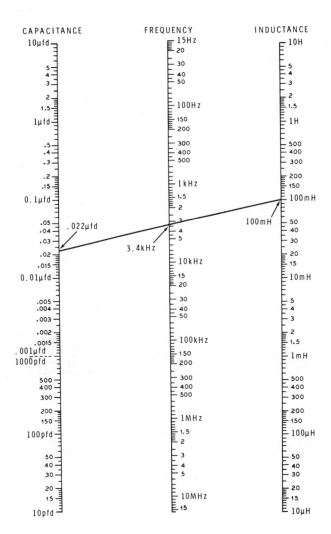

Figure A-1
Resonance Nomograph

Announcing Courses, Lab Manuals, Experimental Parts Packs and Electronic Trainers from Heathkit/Zenith Educational Systems

The book you've been reading is a condensed version of a much larger work from Heathkit/Zenith Educational Systems. If you've enjoyed reading about the fascinating world of electronics, you may want to complement your studies with one or more of our Individual Learning Programs and accompanying laboratory trainers.

Courses are available for: • DC Electronics • AC Electronics • Semiconductor Devices • Electronic Circuits • Test Equipment • Electronic Communications • Digital Techniques • Microprocessors

These courses add a hands-on learning experience to the concepts in this book.

For individuals studying at home, our Individual Learning

Courses are based on logical objectives and include reviews and quizzes to help judge progress. All needed experiment parts are included so you can gain actual electronic experience on a specially designed low-cost trainer.

We also offer 3-part classroom courses that include a Student Text, a Workbook plus Instructor's Guide. The Text establishes competency-based progress while the Workbook has instructions and an electronic Parts Pack for hands-on experiments on our laboratory trainers. The Instructor's Guide offers detailed suggestions to help save class time.

For more information, send for our free catalog. Or see all of our courses and trainers at a Heathkit® Electronic Center near you. Check the white pages of your phone book for locations.

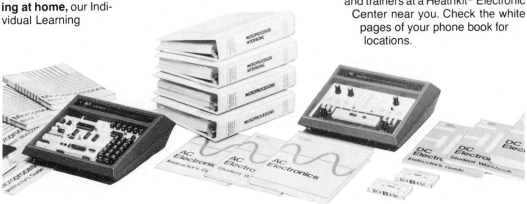